Pro Data Mashup for Power BI

Powering Up with Power Query and the M Language to Find, Load, and Transform Data

Adam Aspin

Apress®

Pro Data Mashup for Power BI: Powering Up with Power Query and the M Language to Find, Load, and Transform Data

Adam Aspin
STAFFORD, UK

ISBN-13 (pbk): 978-1-4842-8577-0   ISBN-13 (electronic): 978-1-4842-8578-7
https://doi.org/10.1007/978-1-4842-8578-7

Managing Director, Apress Media LLC: Welmoed Spahr
Acquisitions Editor: Jonathan Gennick
Development Editor: Laura Berendson
Coordinating Editor: Jill Balzano

Cover photo by Christopher Burns on Unsplash

Distributed to the book trade worldwide by Springer Science+Business Media New York, 1 New York Plaza, Suite 4600, New York, NY 10004-1562, USA. Phone 1-800-SPRINGER, fax (201) 348-4505, e-mail orders-ny@springer-sbm.com, or visit www.springeronline.com. Apress Media, LLC is a California LLC and the sole member (owner) is Springer Science + Business Media Finance Inc (SSBM Finance Inc). SSBM Finance Inc is a **Delaware** corporation.

For information on translations, please e-mail booktranslations@springernature.com; for reprint, paperback, or audio rights, please e-mail bookpermissions@springernature.com.

Apress titles may be purchased in bulk for academic, corporate, or promotional use. eBook versions and licenses are also available for most titles. For more information, reference our Print and eBook Bulk Sales web page at http://www.apress.com/bulk-sales.

Any source code or other supplementary material referenced by the author in this book is available to readers on GitHub via the book's product page, located at www.apress.com. For more detailed information, please visit http://www.apress.com/source-code.

Printed on acid-free paper

Table of Contents

About the Author

Adam Aspin is an independent Business Intelligence consultant based in the United Kingdom. He has worked with SQL Server for over 25 years. During this time, he has developed several dozen reporting and analytical systems based on the Microsoft data and analytics product suite.

A graduate of Oxford University, Adam began his career in publishing before moving into IT. Databases soon became a passion, and his experience in this arena ranges from dBase to Oracle, and Access to MySQL, with occasional sorties into the world of DB2. He is, however, most at home in the Microsoft universe when using the Microsoft data platform—both in Azure and on-premises.

Business Intelligence has been Adam's principal focus for the last 20 years. He has applied his skills for a range of clients in finance, banking, utilities, leisure, luxury goods, and pharmaceuticals. Adam is a frequent contributor to SQLServerCentral. com and Simple Talk. He is a regular speaker at events such as Power BI user groups, SQL Saturdays, and SQLBits. A fluent French speaker, Adam has worked in France and Switzerland for many years.

Adam is the author of *SQL Server 2012 Data Integration Recipes*; *Business Intelligence with SQL Server Reporting Services*; *High Impact Data Visualization in Excel with Power View, 3D Maps, Get & Transform and Power BI*; *Data Mashup with Microsoft Excel Using Power Query and M*; and *Pro Power BI Theme Creation*—all with Apress.

About the Technical Reviewer

 **Ed Freeman** has been a data engineer ever since graduating with a mathematics degree at UCL in 2017. Throughout his career, he has been implementing intelligent cloud and data platforms built on Azure for clients of all sizes and industries. Ed is particularly passionate about Power BI and Microsoft data platform services as a whole edits *Power BI Weekly*, a free weekly newsletter to help the community keep up with the latest and greatest developments from the Power BI ecosystem.

Acknowledgments

Writing a technical book can prove to be a daunting challenge. So I am all the more grateful for all the help and encouragement that I have received from so many friends and colleagues.

First, my heartfelt thanks go, once again, to Jonathan Gennick, the commissioning editor of this book. Throughout the publication process, Jonathan has been an exemplary mentor. He has shared his knowledge and experience selflessly and courteously and provided much valuable guidance.

My deepest gratitude goes yet again to the Apress coordinating team for managing this volume through the rocks and shoals of the publication process. Particular thanks go to Silembarasan Panneerselvam for making a potentially stress-filled trek into a pleasant and stress-free journey.

When delving into the arcane depths of technical products, it is all too easy to lose sight of the main objectives of a book. Fortunately, my good friend Ed Freeman, the technical reviewer, has worked unstintingly to help me retain focus on the objectives of this book. He has also shared his considerable experience of Power Query and the Microsoft Data Platform in the enterprise and has helped me immensely with his comments and suggestions.

Finally, my deepest gratitude has to be reserved for the two people who have given the most to this book. They are my wife and son, who have always encouraged me to persevere while providing all the support and encouragement that anyone could want. I am very lucky to have both of them.

Introduction

Analytics has become one of the buzzwords that define an age. Managers want their staff to deliver meaningful insight in seconds; users just want to do their jobs quickly and well. Everyone wants to produce clear, telling, and accurate analysis with tools that are intuitive and easy to use.

Microsoft recognized these trends and needs a few short years ago when they extended Excel with an add-in called Power Query. Once a mere optional extension to the world's leading spreadsheet, Power Query is now an integral part of Power BI–Microsoft's world-beating analytics application. It allows a user to take data from a wide range of sources and transform them into the base data that they can build on to add metrics, instant analyses, and KPIs to project their insights in eye-catching dashboards.

With Power Query, the era of self-service data access and transformation has finally arrived.

What Is Power Query?

Power Query is a tool that is used to carry out ETL. This acronym stands for **E**xtract, **T**ransform, **L**oad. This is the sequential process that covers:

- Connecting to source data and accessing all or part of the data that you need to bring into Power BI Desktop and Power Query. This is the *extract* phase of ETL.

- Reshaping the data (the "data mashup" process) so that the resulting data is in a form that can be used by Power BI Desktop. Essentially this means ensuring that the data is in a coherent, structured, and complete tabular format. This is the *transform* phase of ETL.

- Returning the data into a Power BI Desktop data model. This is the *load* phase of ETL.

These three phases make up the data ingestion process. So it is worth taking a short look at what makes up each one of them.

Connecting to Source Data

Gone are the days when you manually entered all the data you needed into a spreadsheet. Today's data are available in a multitude of locations and formats and are too voluminous to re-key.

This is where Power Query's ability to connect instantly to 40-odd standard data sources is simply invaluable. Is your accounting data in MS Dynamics? Just connect. Is your CRM data in Salesforce? Just connect. Is your organization using a Data Lake?... You can guess the reply.

Yet this is only a small part of what Power Query can do to help simplify your analyses. For not only can it connect to a multitude of data sources (many of which are outlined in Chapters 1 through 6), it does this via a unified interface that makes connecting to data sources brilliantly simple. On top of this, you can use Power Query to preview the source data and ensure that you are loading exactly what you need. Finally, to top it all, the same interface is used for just about all of the available source data connections. This means that once you have learned to set up *one* connection, you have learned how to *connect to virtually all of the available data sources.*

Data Transformation

Once you have established a connection to a data source, you may need to tweak the data in some way. Indeed, you may even need to reshape it entirely. This is the data mashup process, and it is the area where Power Query shines.

Power Query can carry out the simplest data transformation tasks to the most complex data restructuring challenges in a few clicks. You can:

- Filter source data so that you only load exactly the rows and columns you need

- Extend the source data with calculations or data extracted from existing columns of data

- Cleanse and rationalize the data easily and quickly in a multitude of ways

- Join or split source tables to prepare a logical set of data tables for each specific analytical requirement

- Group and aggregate source data to reduce the quantity of data loaded into Excel

- Prepare source data tables to become a usable data model

This list merely scratches the surface of all that Power Query can do to mash up your data. It is, without hyperbole, unbelievably powerful at transforming source data. Indeed, it can carry out data ingestion and transformation tasks that used to be the preserve of expensive products that required complex programming skills and powerful servers.

All of this can now be done using a code-free interface that assists you in taking the messiest source data and delivering it to Excel as limpid tables of information ready to work with.

The aim of this short book is to introduce the reader to this brave new world of user-driven data integration. This will not involve a complete tour of Power BI Desktop, however. The product is simply too vast for that. Consequently, this book concentrates on data mashup using Power BI Desktop and Power Query. If you need to learn further aspects of the Power BI Desktop ecosystem, then two companion volumes are available:

- *Pro DAX and Data Modeling in Power BI*

- *Pro Power BI Dashboard Creation*

The first explains how to meld disparate data sources into a unified Power BI data model that you then extend using DAX—the built-in analytics language. The second guides you through the process of dashboard creation using all the range of available visuals and techniques.

Although a basic knowledge of the MS Office suite will help, this book presumes that you have little or no knowledge of Power BI Desktop. This product is therefore explained from the ground up with the aim of providing the most complete coverage possible of the way data can be discovered and loaded into Power BI as the basis for user-driven dashboards. Hopefully, if you read the book and follow the examples given, you will arrive at a level of practical knowledge and confidence that you can subsequently apply to your own data ingestion requirements. This book should prove invaluable to business intelligence developers, MS Office power users, IT managers, and finance experts—indeed anyone who wants to deliver efficient and practical business intelligence to their colleagues. Whether your aim is to develop a proof of concept or to deliver a fully fledged BI system, this book can, hopefully, be your guide and mentor when it comes to ensuring that the data you base your analytics on is clean and well structured.

If you wish, you can read this book from start to finish as it is designed to be a progressive tutorial that will help you to learn Power Query. However, as Power Query is composed of four main areas, this book is broken down into four sets of chapters that focus on the various key areas of the product. It follows that you can, if you prefer, focus on individual topics in Power Query without having to take a linear approach to reading this book.

- Chapters 1 through 6 show you how to connect to a range of varied data sources and bring this data into Excel using Power Query. Depending on the source data that you need to use, you may only need to dip into parts of these chapters to find guidance on how to use a specific source data type.

- Chapters 7 through 11 explain how to transform and clean data so that you can use it for analysis. These data transformations range from the extremely simple to the potentially complex. Indeed, they are as potentially vast as data itself. You may never need to apply all of the extensive range of data modification and cleansing techniques that Power Query can deliver, but just about everything that it can do is explained in detail in these chapters.

- Chapters 12 and 13 explain how to tame the real world of data loading and transformation. Here, you will learn how to organize and manage your queries, as well as how to add parameters to make them more interactive and resilient.

- Chapter 14 introduces you to M—the language that Power Query uses to transform your data. Using M, you can push your data ingestion and transformation routines to new heights that are simply not possible using just the Power Query interface.

Inevitably, not every question can be answered, and not every issue can be resolved in one book. I truly hope that I have answered many of the essential Power Query questions that you will face when ingesting data into Power BI Desktop. Equally I hope that I have provided ways of solving a reasonable number of the challenges that you may encounter. I wish you good luck in using Power Query and Power BI Desktop to prepare and deliver your insights. And I sincerely hope that you have as much fun with it as I had writing this book.

CHAPTER 1

Discovering and Loading Data with Power BI Desktop

Before you can use Power BI Desktop to present any analysis or discover new insights, you need data. Your sources could be in many places and in many formats. Nonetheless, you need to access them, look at them, select them, and quite possibly restructure them or clean them up to some extent. You may also need to join many separate data sources before you shape the data into a coherent model that you can use as the foundation for your dashboards and reports. The amazing thing is that you can do all of this using Power BI Desktop without needing any other tools or utilities.

Discovering, loading, cleaning, and modifying source data are some of the many areas where Power BI Desktop really shines. It allows you to accomplish the following:

- *Data discovery*: Find and connect to a myriad of data sources containing potentially useful data. This can be from both public and private data sources. This is the subject of Chapters 1 through 6.

- *Data loading*: Select the data you have examined and load it into Power BI Desktop for shaping. This, too, is handled in Chapters 1 through 6.

- *Data modification*: Modify the structure of each table that you have imported and then filter and clean the data itself (we will look at this in detail in Chapters 7 through 13).

- *Programming data transformations*: This is described in Chapter 14.

© Adam Aspin 2022
A. Aspin, *Pro Data Mashup for Power BI*, https://doi.org/10.1007/978-1-4842-8578-7_1

- *Data shaping*: Join tables to create a clear, unified, and accessible data model. As this is a fairly complex subject, it is covered in the companion volume "Pro Dax and Data Modeling with Power BI Desktop".

Although I have outlined these five steps as if they are completely separate and sequential, the reality is that they often blend into a single process. Indeed, there could be many occasions when you will examine the data *after* it has been loaded into Power BI Desktop—or clean data *before* you load it. The core objective will, however, always remain the same: find some data and then load it into Power BI Desktop where you can tweak, clean, and shape it.

This process could be described simplistically as "First, catch your data." In the world of data warehousing, the specialists call it ETL, which is short for **E**xtract, **T**ransform, and **L**oad. Indeed, it is also known as ELT (**E**xtract, **L**oad, and **T**ransform), which can also be handled by Power BI Desktop. Despite the reassuring confidence that the acronym brings, this process is rarely a smooth, logical progression through a clear-cut series of steps. The reality is often far messier. You may often find yourself importing some data, cleaning it, importing some more data from another source, combining the second table with the first one, removing some rows and columns, and then repeating these operations, as well as many others, several times over.

In this book, I will show you how the process can work in practice using Power BI Desktop. I hope that this will make the various steps that comprise an ETL process clearer. All I am asking is that you remain aware that the range of options that Power BI Desktop includes make it a multifaceted and tremendously capable tool. The science is to know *which* options to use. The art is to know *when* to use them.

The Data Load Process

Let's begin with a rapid overview of what you need to do to get some data into Power BI Desktop. This requires you to follow the download instructions in Appendix A, to prepare the source files in the `C:\PowerBIDesktopSamples` folder.

Once you have launched Power BI Desktop, you are faced with the splash screen that looks something like the one that you can see in Figure 1-1.

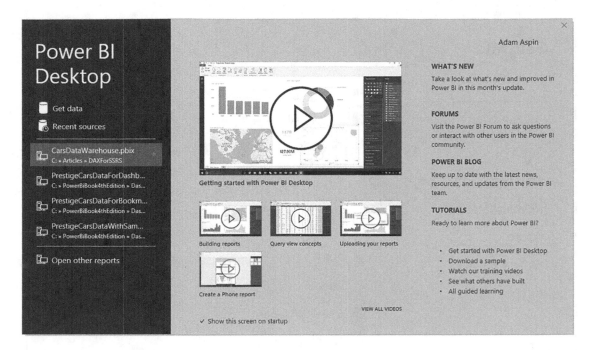

Figure 1-1. *The splash screen*

Given that you are working with an application that lives and breathes data, it is not really surprising that the first step in a new analytical challenge is to find and load some data. So the following explains what you have to do (assuming that you have downloaded the sample data that accompanies this book from the Apress website—as is explained in Appendix A):

1. Click Get data in the splash screen. The Get Data dialog will appear, as shown in Figure 1-2.

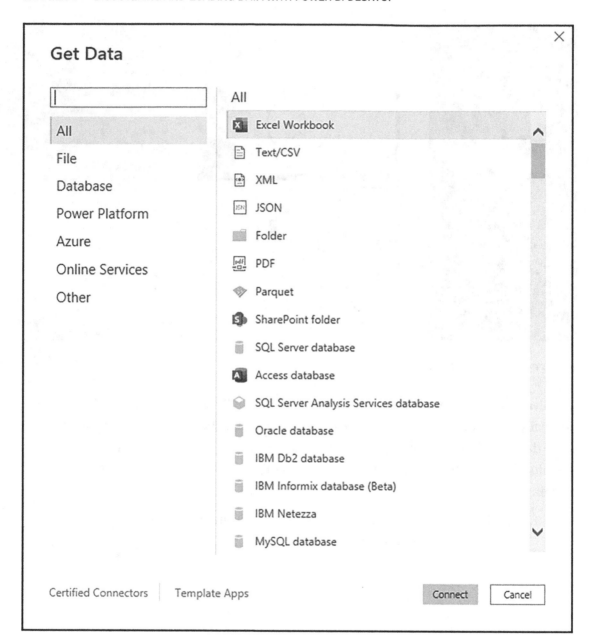

Figure 1-2. *The Get Data dialog*

2. In the list of all the possible data sources on the right of this dialog,
    select Excel Workbook, and then click Connect. The Windows
    Open File dialog will appear.

3. Click the file C:\PowerBiDesktopSamples\
    BrilliantBritishCars.xlsx. The Windows Open dialog will look
    like the one in Figure 1-3.

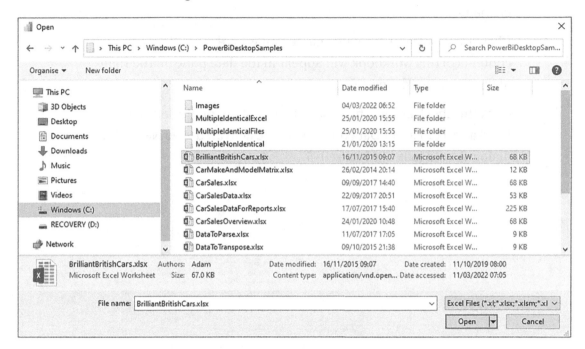

Figure 1-3. *The Windows Open File dialog when loading data from a file source*

4. Click the Open button. The Connecting dialog will appear for a
    second or two (it looks like Figure 1-4), and then the Navigator
    dialog will appear.

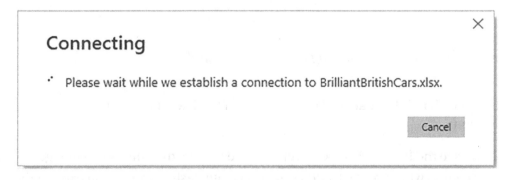

Figure 1-4. *The Connecting dialog*

5. You will see that the BrilliantBritishCars.xlsx file appears on
    the left of the Navigator dialog and that any workbooks, named
    ranges, or data tables that it contains are listed under the
    file name.

6. Click the BaseData worksheet name that is on the left. The
    contents of this workbook will appear in the data pane on the right
    of the Navigator dialog.

7. Click the check box for the BaseData worksheet on the left. The
    Load and Transform Data buttons will be activated. The Navigator
    dialog should look like Figure 1-5.

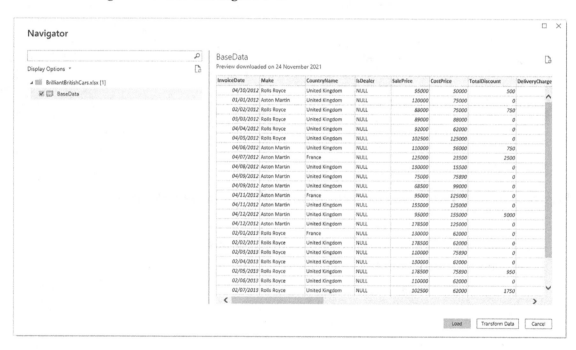

Figure 1-5. *The Navigator dialog with data selected*

8. Click Load. The data will be loaded from the Excel file into Power
    BI Desktop.

You will see the Power BI Desktop report window, like the one shown in Figure 1-6.
This is the canvas where you will add visuals to create dashboards. If you expand the
BaseData table in the Fields pane on the right of the Power BI Desktop application, you
can see all the columns from the Excel worksheet are now fields.

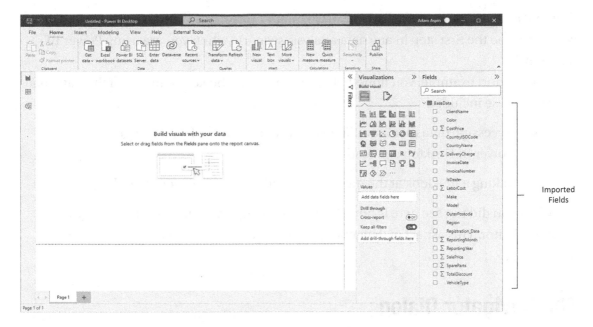

Figure 1-6. *Data available in Power BI Desktop*

I imagine that loading this data took under a minute. Yet you now have a fully operational data model in Power BI Desktop that is ready to feed data into your dashboard.

So if you are itching to race ahead and actually create a couple of tables, charts, and visuals, then you can always use the companion volume *Pro Power BI Dashboard Creation* to guide you through the process of building charts and dashboards.

However, as this book is all about getting and transforming data in Power BI Desktop, I would like to pause for an instant and explain exactly what you have seen so far.

Note   The Get Data dialog contained a reference to certified connectors. These are developed by third parties but certified and distributed by Microsoft.

Understanding Data Load

What you have seen so far is an extremely rapid dash through a Power BI Desktop data load scenario. In reality, this process can range from the blindingly simple (as you just saw) to the more complex where you join, filter, and modify multiple tables from different sources (as you will discover in Chapters 7 through 14). However, loading data will always be the first step in any data analysis scenario when you are using Power BI Desktop.

In this short example, you nonetheless saw many of the key elements of the data load process. These included

- Accessing data that is available in any of the source formats that Power BI Desktop can read

- Taking a first look at the data before loading it into Power BI Desktop

What you did not see here is how Power BI Desktop can transform the source data in Query Editor. This aspect of data manipulation is covered extensively in Chapters 7 through 12.

The Navigator Dialog

One key aspect of the data load process is using the Navigator dialog correctly. You saw this dialog in Figure 1-5. The Navigator window appears when connecting to many, but not all, data sources. It allows you to

- Take a quick look at the available data tables in the data source

- Look at the data in individual tables

- Select one or more data tables to load into Power BI Desktop

---

Note   If you want to display more sample data, you can increase the height and width of the Navigator dialog. Alternatively, you can scroll through the window on the right of the Navigator.

Depending on the data source to which you have connected, you might see only a few data tables in the Navigator window, or hundreds of them. In any case, what you can see are the structured tables that Power BI Desktop can recognize and is confident that it can import. Equally dependent on the data source is the level of complexity of what you will see in the Navigator window. If you are looking at a database server, for instance, then you may start out with a list of databases and you may need to dig deeper into the arborescence of the data by expanding databases to list the available data tables and views.

The more you work with Power BI Desktop, the more you will use the Navigator dialog. So it seems appropriate to explain at this early juncture some of the tricks and techniques that you can apply to make your life easier when delving into potential sources of data.

Let's start by taking a closer look at the available options. These are highlighted in Figure 1-7.

Figure 1-7. *The Navigator dialog*

The Navigator dialog is essentially in two parts:

- *On the left*: The hierarchy of available data sources. These can consist of a single table or multiple tables, possibly organized into one or many folders.

- *On the right*: A preview of the data in the selected element.

The various Navigator dialog options are explained in the following sections.

Searching for Usable Data

You will, inevitably, come across cases where the data source that you are connecting to will contain hundreds of potential datasets that you could use. These datasets are generally referred to as tables. This is especially true for databases. Fortunately, Power BI Desktop lets you filter the tables that are displayed extremely easily.

1. In the Navigator dialog, click inside the Search box.

2. Enter a part of a table name that you want to isolate.

3. Click the magnifying glass icon at the right of the Search box. The list of tables will be filtered to show only tables containing the text that you entered. You can see an example of this in Figure 1-8.

Figure 1-8. *Table search in the Navigator dialog*

Once you have previewed and selected the tables that you want to use, simply click the cross at the right of the Search box. Navigator will clear the filter and display all the tables in the data source.

Display Options

Clicking Display Options in the Navigator dialog will show a pop-up menu with two options:

- Only selected items
- Enable data previews

Only Selected Items

Selecting this option will prevent any tables that you have not selected from appearing in the data source pane. This is not selected by default.

Enable Data Previews

Selecting this option will show a small subset of the data available in the selected table. You could choose to disable data previews if the connection to the source data is slow.

Refresh

If you need to, you can refresh either or both of the following:

- The data sources
- The data preview

Source Data Refresh

Clicking the preview button under the search bar will refresh the source data in the source data pane.

Data Preview Refresh

Clicking the preview button on the top right of the Navigator dialog will refresh the preview data visible on the right. This may take a few seconds–or more depending on the data source.

Select Related Tables

Clicking the Select Related Tables button is only valid for database sources, such as Microsoft SQL Server or Oracle. If the source database has been designed correctly to include joins between tables, then this option will automatically select all tables that are linked to any tables that you have already selected.

Note   In a database source, some tables can be related to other tables that are, themselves, related to other tables. This hierarchy of connections is not discovered in its entirety when you click Select Related Tables. In other words, you might have to select several tables and click this button repeatedly to select *all* the tables that you need.

You will see much, much more of the Navigator window in the following three chapters.

The Navigator Data Preview

The Navigator Data Preview pane (on the right) is, as its name implies, a preview of the data in a data source. It provides a brief overview of the *top few records* in any of the tables that you want to look at. Given that the data you are previewing could be hundreds of columns wide and hundreds of rows deep, there could be scroll bars for the data table visible inside the Navigator Data Preview. Remember, however, that you are not examining *all* the available data and are only seeing a *small sample of the available records.*

Power BI Desktop can preview and load data from several different sources. Indeed (as you can see from the list of possible data sources in the Get Data dialog), it can read most of the commonly available enterprise data sources as well as many, many others. What is important to appreciate is that Power BI Desktop applies a common interface to the art and science of loading data, whatever the source. So whether you are examining a SQL Server or an Oracle database, an XML file or a text file, a web page or a big data source, you will always be using a standardized approach to examining and loading the data. This makes the Power BI Desktop data experience infinitely simpler—and extremely reassuring. It means that you spend less time worrying about technical aspects of data sources and you are free to focus on the data itself.

Note   The Navigator Data Preview is a brilliant data discovery tool. Without having to load any data, you can take a quick look at the data source and any data that it contains that can (probably) be loaded by Power BI Desktop. You can then decide if it is worth loading. This way, you do not waste time on a data load that could be superfluous to your needs.

Modifying Data

Once you have one or more data connections (which Power BI Desktop calls queries) that can connect to data sources and bring the data into this environment, you can start thinking about the next step—transforming the data so that it is ready for use. Depending on the number of data sources that you are handling and the extent of any modifications that are required, this could vary from the simple to the complex. To give a process some structure, I advise that you try to break down any steps into the following main threads:

- *Structure the data*: This covers filtering out records to reduce the size of the table, as well as removing any extraneous columns. It may also involve adding columns that you create by splitting existing columns, creating calculated columns, or even joining queries.

- *Cleanse and modify the data*: This is also known as *data transformation* (the *T* in ETL). It encompasses the process of converting text data to uppercase and lowercase, as well as (for instance) removing nonprinting characters. Rounding numbers and extracting date parts from date data are also possible (among the many dozens of other available transformations).

For the moment, however, it is only important to understand that Power BI Desktop can do all of this if you need it to. Transforming data is explained in detail in Chapters 7 through 14.

Data Sources

Previously in this chapter, you saw how quickly and easily you can load data into Power BI Desktop. It is now time to take a wider look at the *sources* of data that Power BI Desktop can ingest and manipulate.

As the sheer wealth of possible data sources can seem overwhelming at first, Power BI Desktop groups potential data sources into the following categories:

- *File*: Includes Excel files, CSV (comma-separated values) files, text files, JSON files, and XML files. Power BI Desktop can even load entire folders full of files. You will discover many of these in the following chapter.

- *Database*: A comprehensive collection of relational databases that are currently in the workplace and in the cloud, including (among others) MS Access, SQL Server, and Oracle. The full list of those available when this book went to press is given in Chapter 3.

- *Power Platform*: These sources cover data that is made available in the Microsoft Power Platform and the Power BI Service. You can see how some of these are used with Power BI Desktop in Chapter 5.

- *Azure*: This option lets you see an immense range of data source types that are hosted in Azure–the Microsoft cloud. This covers data formats from SQL Server through to big data sources. You can see how a few of these are used with Power BI Desktop in Chapter 5.

- *Online Services*: These sources range from SharePoint lists to Salesforce, Dynamics 365 to Google Analytics—and many, many others. A few of these are examined in Chapter 5.

- *Other*: A considerable and ever-growing range of data sources, from SharePoint lists to Microsoft Exchange. Some of these will be touched on in Chapter 6 of this book.

> **Note** Power BI can only load *tabular* data. This means (as you will discover in Chapters 2 through 6) that the source data must be a structured set of rows and columns.

The list of possible data sources is changing all the time, and you need to be aware that you have to look at the version of Power BI Desktop that you are using if you want an exhaustive list of all the available data sources that you can use. Indeed, I expect that several more will have been added by the time that you read this book.

You can also list the contents of folders on any available local disk, network share, or even in the cloud and then leverage this to import several files at once. Similarly (if you have the necessary permissions), you can list the databases and data available on the database servers you connect to. This way, Power BI Desktop can provide not only the data but also the *metadata*—or data about data—that can help you to take a quick look at potential sources of data and only choose those that you really need.

Unfortunately, the sheer range of data sources from which Power BI Desktop can read data is such that we do not have space in a few chapters to examine the minutiae of every one. Consequently, we will take a rapid tour of *some* of the most frequently used data sources in this and the next few chapters. Fortunately, most of the data sources that Power BI Desktop can read are used in a similar way. This is because the Power BI Desktop interface does a wonderful job of making the arcane connection details as unobtrusive as possible. So even if you are faced with a data source that is not described in these chapters, you will nonetheless see a variety of techniques that can be applied to virtually any of the data sources that Power BI Desktop can connect to.

> **Note** The list of data sources that Power BI Desktop can access is growing all the time. Consequently, when you read this book, you will probably find even more sources than those described in this and the first six chapters of this book.

The Power BI Desktop Screen

Getting data into Power BI Desktop is so vital that even the Power BI Desktop initial screen reminds you that you need data to start working. If you close the splash screen that you saw in Figure 1-1, you will see not a blank canvas, but images to encourage you to start adding data straight away. You can see this in Figure 1-9.

Figure 1-9. *The Power BI Desktop initial screen*

If you prefer, you can load data from Excel (or even one of the other three suggested data sources) simply by clicking one of the tiles that are visible on the desktop canvas. Alternatively, you can display the Get Data dialog by clicking Get data from another source.

The Get Data Dialog

One of the major strengths of Power BI is that it allows you to access hundreds of different data sources that you can then use to create dashboards. Unless you are using one of a handful of extremely common data sources (such as Excel), you will need to begin the process by indicating to Power BI Desktop what type of data source you want to connect to.

Accessing most data sources is handled by the Get Data dialog. You can display this by

- Clicking Get data in the splash screen

- Clicking Get data from another source in a blank dashboard canvas

- Clicking the Get data button followed by More in the Home ribbon

You saw the Get Data dialog in Figure 1-2.

Given the plethora of potential data sources, Power BI Desktop groups them into the following categories to make locating a data source easier. You can click on any of the sections on the left to see a filtered list of the selected type of data sources on the right.

The next five chapters will guide you through the data load process for many of the most common data sources that you may need to use when working with Power BI.

As there are so many potential data sources to choose from, the Get Data dialog also has a simple search facility. All you have to do is to click inside the search box at the top left and enter part of the name of the data source that you are looking for. You can see an example of this in Figure 1-10 where the Get Data dialog is showing data sources that contain the characters "pos".

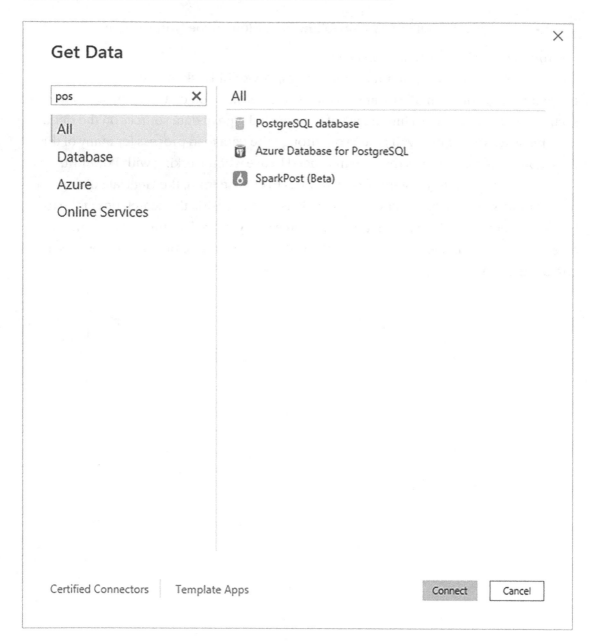

Figure 1-10. *Searching the Get Data dialog for a data source*

Clicking the small cross icon at the right of the search box will clear the search box and display all data sources.

Conclusion

In this chapter, you have seen how Power BI Desktop can connect to any of a wide range of data sources. You have seen that as long as you know what kind of data you want to load—and that Power BI Desktop has an available connector to this data—you can preview and load the data.

Now it is time to delve deeper into the details of some of the various data sources that you can use with Power BI Desktop. The next chapter will start on your journey by introducing many of the file-based data types that you can use to create analytical dashboards.

CHAPTER 2

Discovering and Loading File-Based Data with Power BI Desktop

File Sources

Sending files across networks and over the Internet or via email has become second nature to most of us. As long as the files that you have obtained conform to some of the widely recognized standards currently in use (of which you will learn more later), you should have little difficulty loading them into Power BI Desktop.

As the first part of your journey through the data mashup process, this chapter will show you how to find and load data from a variety of file-based sources. These kinds of data are typically those that you can either locate on a shared network drive, download from the Internet, receive as an email attachment, or copy to your computer's local drive. The files that are used in the examples in this chapter are available on the Apress website. If you have followed the download instructions in Appendix A, then these files will be in the `C:\PowerBIDesktopSamples` folder.

The file sources that Power BI Desktop can currently read and from which it can load data are given in Table 2-1.

© Adam Aspin 2022
A. Aspin, *Pro Data Mashup for Power BI*, https://doi.org/10.1007/978-1-4842-8578-7_2

Table 2-1. *File Sources*

File Source	Comments
Excel	Allows you to read Microsoft Excel files (versions 97 to 2021) and load worksheets, named ranges, and tables
Text/CSV	Lets you load text files that conform to the CSV (comma-separated values) format as well as text files using a variety of column separators
XML	Allows you to load data from XML files
Parquet	Lets you read files in the Parquet format
Folder	Lets you load the information about all the files in a folder
SharePoint folder	Allows you to list the files in a SharePoint folder
Access database	Lets you connect to a Microsoft Access file on your network and load queries and tables
JSON	Helps you to load data from JSON files
PDF	Allows you to preview PDF files and load data from any tables that are in the document

In this chapter, we will be looking at how to import data from

- CSV files

- Text files

- XML files

- Excel files

- Access databases

- PDF files

More advanced techniques (such as importing the contents of entire folders of text or Excel files or importing complex XML files and JSON files) are described in Chapter 10. I prefer to handle these separately as they require more advanced knowledge of data transformation techniques—and you need to learn these first.

> **Note**   I realize that Power BI Desktop considers MS Access to be a database and
> not a "file" data type. While I completely agree with this classification, I prefer
> nonetheless to treat Access as if it were a file-based data source, given that all the
> data resides in a single file that can be copied and emailed, and not in a database
> on a distant server. For this reason, we will look at MS Access in this chapter, and
> not the next one that deals with corporate databases.

Loading Data from Files

It is time to start looking at the heavy-lifting aspect of Power BI Desktop and how you can
use it to load data from a variety of different sources. I will begin on the bunny slopes (or
"nursery" slopes as we say in the UK) with a simple example of loading data from a text
file. Then, given the plethora of available data sources, and to give the process a clearer
structure, we will load data from several of the ubiquitous file-based data sources that
are found in most workplaces. These data sources are the basis of the data that you will
learn to tweak and "mash up" in Chapters 7 through 12. This data can then become the
basis of the dashboards that you will create.

CSV Files

The scenario is as follows: you have been given a CSV file containing a list of data.
You now want to load this into Power BI Desktop so that you can look at the data and
consider what needs to be done (if anything) to make it usable.

First, you need an idea of the data that you want to load. If you open the source file
C:\PowerBIDesktopSamples\Countries.csv with a text editor, such as Notepad, you can
view its contents. This is what you can see in Figure 2-1.

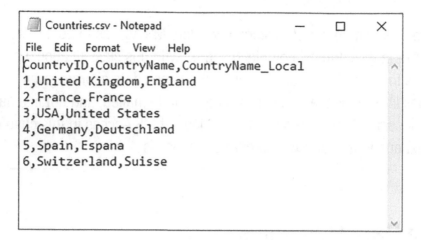

Figure 2-1. *The contents of the Countries.csv file*

The following steps explain what you have to do to load the contents of this file into Power BI Desktop:

1. Open Power BI Desktop and close the splash screen.

2. In the Power BI Desktop Home ribbon, click the Get Data button (and not the small triangle that displays menu options). The Get Data dialog will appear.

3. Click File on the left. You will see something like Figure 2-2 (the Get Data dialog).

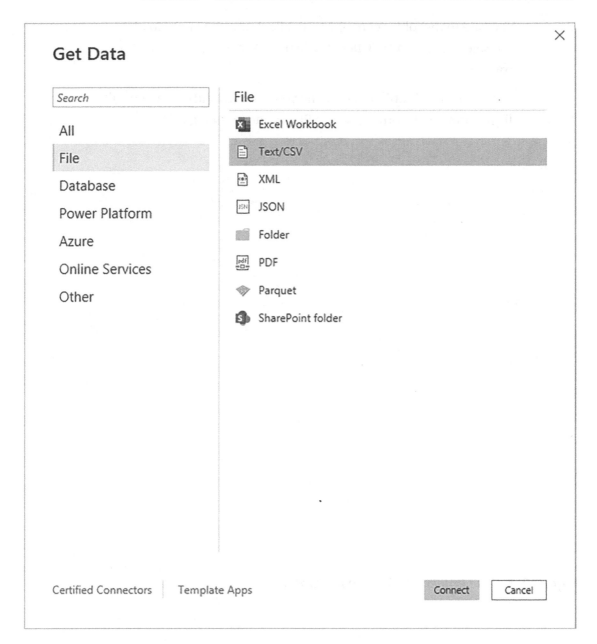

Figure 2-2. *The file data connectors in Power BI Desktop*

4. Click Text/CSV on the right of the dialog.

5. Click Connect (or double-click Text/CSV). The Open dialog will appear.

6. Navigate to the folder containing the file that you want to load
    and select it (C:\PowerBIDesktopSamples\Countries.csv in this
    example).

7. Click Open (or double-click the file you want to load). A dialog will
    display the initial contents of the file, as shown in Figure 2-3.

Figure 2-3. *The Power BI Desktop file dialog*

8.   Click the Transform Data button. The Power Query window
     appears; it contains a sample of the contents of the CSV file—or
     possibly the entire file if it is not too large. You can see this in
     Figure 2-4.

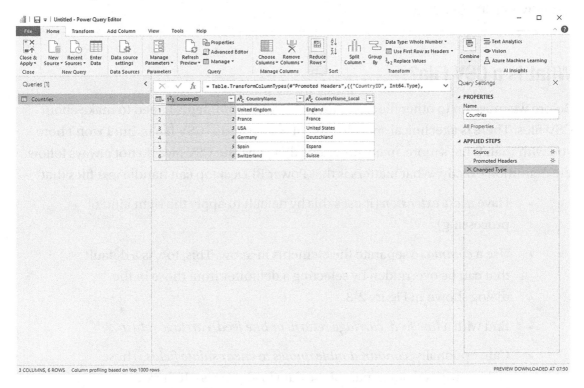

Figure 2-4. *The Power Query window with the contents of a CSV file loaded*

9.   Click the Close & Apply button (or Close & Apply in the pop-up
     menu for the Close & Apply button) in the Power Query window
     (you can see this at the top left of Figure 2-4). Power Query will
     close and return the focus to the Power BI Desktop window, where
     you can see that the Countries table appears in the Fields list on
     the right of the screen.

And that, for the moment, is that. You have loaded the file into Power BI Desktop in a
matter of a few clicks, and it is ready for use in dashboards and reports. In later chapters,
you will learn how to shape this data. For the moment, however, let's continue looking at
some other file-based data sources.

Note   Power Query is where you transform data. Although it is not used in the first six chapters of this book, I prefer to introduce it now, as taking a detour via Power Query and then returning to the Power BI Desktop window is the same as simply loading the data.

What Is a CSV File?

Before we move on to other file types, there are a few comments I need to make about CSV files. There is a technical specification of what a "true" CSV file is, but I won't bore you with that. What's more, many programs that generate CSV files do not always follow the definition exactly. What matters is that Power BI Desktop can handle text files that

- Have a *.csv extension* (it uses this by default to apply the right kind of processing).

- Use a *comma* to separate the elements in a row. This, too, is a default that can be overridden by selecting a delimiter from those in the dialog shown in Figure 2-3.

- End with a *line feed, carriage return, or line feed/carriage return*.

- Can, optionally, *contain double quotes to encapsulate fields*. These will be stripped out as part of the data load process. If there are double quotes, they do not have to appear for every field nor even for every record in a field that can have occasionally inconsistent double quotes.

- Can contain "irregular" records, that is, rows that do not have every element found in a standard record. However, the first row (whether or not it contains headings) must cover every element found in all the remaining records in the list. Put simply, any other record can be shorter than the first one but cannot be longer.

- Do not contain anything other than the data itself. If the file contains header rows or footer rows that are not part of the data, then Power BI Desktop cannot load the table without further work. There are workarounds to this all-too-frequent problem; one is given in Chapter 8.

Note   Another way of accessing CSV files is to click Get Data and select Text/CSV in the pop-up menu.

Text Files

If you followed the process for loading a CSV file in the previous section, then you will find that loading a text file is virtually identical. This is not surprising. Both are text files, and both should contain a single list of data. The following are the core differences:

- A text file can have something *other* than a comma to separate the elements in a list. You can specify the delimiter when defining the load step.

- A text file should normally have the extension .txt (though this, too, can be overridden).

- A text file *must* be perfectly formed; that is, every record (row) must have the same number of elements as every other record.

- A text file, too, *must not* contain anything other than the table if you want a flawless data load the first time.

- If a text file encounters difficulties, it should import the data as a single column that you can then try and split up into multiple columns, as described in Chapter 8.

Here, then, is how to load a text file into Power BI Desktop:

- In the Power BI Desktop ribbon, click Get Data ➤ Text/CSV. The Open dialog will be displayed.

- Navigate to the folder containing the file and select the file (`C:\PowerBIDesktopSamples\CountryList.txt` in this example).

- Click Open. A dialog will display the initial contents of the file (this dialog is essentially identical to the one that you saw for CSV files in Figure 2-3). You can, of course, double-click the file name rather than click Open.

- Click the Cancel button (because after a quick look at the contents of the file, you have decided that you do not really need it).

Note   As text-based files (which include CSV files) are such a frequent source of data, you will nearly always see the Text/CSV option directly accessible in the pop-up menu that you access by clicking the small triangle in the Get Data button in the Home ribbon. If this option is not visible, you can instead select Get Data ➤ File and select Text/CSV, as you did previously.

Where Power BI Desktop is really clever is that it can make a very educated guess as to how the text file is structured; that is, it can nearly always guess the field separator (the character that isolates each element in a list from the other elements). And so not only will it break the list into columns, but it will also avoid importing the column separator. If it does not guess correctly, then don't despair. You will learn how to correct this in Chapter 8.

Looking at the contents of a file and then deciding not to use it is part and parcel of the *data discovery* process that you will find yourself using when you work with Power BI Desktop. The point of this exercise is to show you how easy it is to glance inside potential data sources and then decide whether to import them into the data model or not. Moreover, it can be easier to see the first few rows of large text or CSV files directly in the Load dialog of Power BI Desktop than it is to open the whole file in a text editor.

Tip   At the risk of stating the obvious, you can press Enter to accept a default choice in a dialog and press Esc to cancel out of the dialog.

Text and CSV Options

You can see in Figure 2-3 that there are few options available that you can tweak when loading text or CSV files. Most of the time, Power BI Desktop will guess the correct settings for you. However, there could be times when you will need to adjust these parameters slightly. The potential options that you can modify are

- File origin

- Delimiter

- Data type detection

File Origin

This option defines the character encoding in which the file is stored. Different character sets can handle differing ranges of characters, such as accents and other diacritics. Normally this information is correctly interpreted by Power BI Desktop, and you should only need to select a different character set (file origin) on very rare occasions.

Delimiter

Power BI Desktop will try and guess the special character that is used in a text or CSV file to separate the "columns" of data. Should you wish to override the chosen delimiter, you have the choice of

- Colon

- Comma

- Equals sign

- Semicolon

- Space

- Tab character

You can also decide to enter a custom delimiter such as the pipe (|) character or even specify that every field has a fixed width. Choosing either of these options will display another entry field where you can type in the required delimiter—or the fixed length that you require.

Data Type Detection

Power BI Desktop will make an educated guess at the data encoding and data type that is used in a text or CSV file. By default, to save time, it will only read the first 200 records. However, you can choose from any of the following three options:

- Read the first 200 rows

- Read the entire file

- No data type detection

Note   Be warned that reading a large file in its entirety can take quite a while. However, without accurate data type detection, you risk seeing some weird characters in the data that you load.

Simple XML Files

XML, or Extensible Markup Language, is a standard means of sending data between IT systems. Consequently, you likely will load an XML file one day. Although an XML file is just text, it is text that has been formatted in a very specific way, as you can see if you ever open an XML file in a text editor such as Notepad. Do the following to load an XML file:

1. In the Power BI Desktop ribbon, click the small triangle on the Get Data button, and then click More in the menu that appears. Next, in the Get Data dialog, select File and XML.

2. Click Connect. The Open dialog will appear.

3. Navigate to the folder containing the file and select the file (C:\ PowerBIDesktopSamples\ColoursTable.xml in this example).

4. Click Open. The Navigator dialog will open.

5. Click the Colours table in the left-hand pane of the Navigator dialog. The contents of this part of the XML file will be displayed on the right of the Navigator dialog, as shown in Figure 2-5.

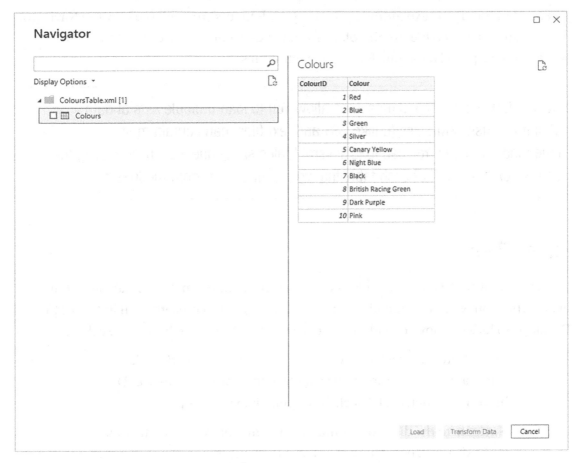

Figure 2-5. *The Navigator dialog before loading an XML file*

6. Click the check box to the left of the Colours table on the left. The Load and Transform buttons will be enabled.

7. Click the Transform Data button. The Power Query window will display the contents of the XML file.

8. Click the Close & Apply button in the Power BI Desktop Data window. You will see that the Colours table appears in the Fields list on the right of the screen.

The actual internal format of an XML file can get extremely complex. Sometimes, an XML file will contain only one table; sometimes, it will contain many separate tables. On other occasions, it will contain one table whose records contain nested levels of data that you need to handle by expanding a hierarchy of elements. You will see how the Navigator dialog handles nested hierarchies of XML data in Chapter 12—once you have learned some of the required data transformation techniques.

Note   Certain types of data source allow you to load multiple sets of data simultaneously. XML files (unlike CSV and text files) can contain multiple independent tables. You can load several tables simultaneously by selecting the check box to the left of each table that you want to load from the XML file.

Excel Files

You are probably already a major Excel user and have many, many spreadsheets full of data that you want to rationalize and use for analysis and presentation in Power BI Desktop. So let's see how to load a couple of worksheets at once from an Excel file.

1. In the Power BI Desktop Home ribbon, click Excel workbook. Alternatively, click the small triangle at the bottom of the Get Data button and then click Excel. The Open dialog will appear.

2. Navigate to the directory containing the file that you want to look at (`C:\PowerBIDesktopSamples` in this example).

3. Select the source file (InvoicesAndInvoiceLines.xlsx in this example) and click Open. The Navigator dialog will appear, showing the worksheets, tables, and ranges in the workbook file.

4. Click one of the tables listed on the left of the Navigator dialog. The top few rows of the selected spreadsheet will appear on the right of the dialog to show you what the data in the chosen table looks like.

5. Click the check boxes to the left of the Invoices and InvoiceLines tables on the left. The Navigator dialog will look like the one shown in Figure 2-6.

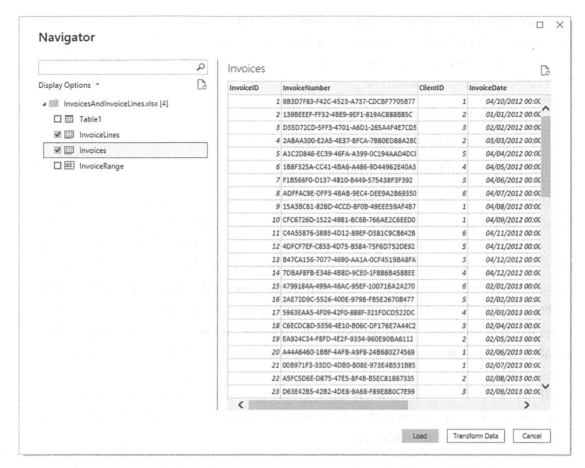

Figure 2-6. *The Navigator dialog before loading data from an Excel workbook*

6. Click Load. The selected worksheets will be loaded into the Power BI Desktop data model and will appear in the Fields list in the Report window.

As you can see from this simple example, having Power BI Desktop read Excel data is really not difficult. You could have edited this data in Power Query before loading it, but as the data seemed clean and ready to use, I preferred to load it straight into Power BI Desktop (or rather the Power BI Desktop data model). As well, you saw that Power BI

Desktop can load multiple tables at the same time from a single data source. However, you might still be wondering about a couple of things that you saw during this process, so here are some anticipatory comments:

The Navigator dialog displays

- Worksheets (Invoices and InvoiceLines in Figure 2-6)

- Named ranges (InvoiceRange in Figure 2-6)

- Named tables (Table1 in Figure 2-6)

Each of these elements is represented by a different icon in the Navigator dialog. Sometimes, these can, in effect, be duplicate references to the same data, so you should really use the most precise data source that you can. For instance, I advise using a named table or a range name rather than a worksheet source, as the latter could easily end up containing "noise" data (i.e., data from outside the rows and columns that interest you), which would make the load process more complex than it really needs to be—or even cause it to fail. Indeed, unless a worksheet is prepared and structured in a simple tabular format, ready for loading into Power BI Desktop, you could end up with superfluous data in your data model.

However, the really cool thing is that you can load as many worksheets, tables, or ranges as you want at the same time from a single Excel workbook. You do not need to load each source table individually.

Note   Power BI Desktop will list and use data connections to external data sources (such as SQL Server, Oracle, or SQL Server Analysis Services) in a source Excel workbook *if* the data connection is active and has returned data to the workbook. Once a link to Power BI Desktop has been established, you can delete the data table itself in the source Excel workbook—and still load the data over the data connection in the source workbook into Power BI Desktop.

Power BI Desktop will *not* take into account any data filters on an Excel data table but will load all the data that is in the source table. Consequently, you will have to reapply any filters (of which you'll learn more in Chapter 8) in Power BI Desktop if you want to subset the source data.

There are a couple of important points that you need to be aware of at this juncture:

- Multiple worksheets, tables, or named ranges can all be imported from the same workbook (i.e., Excel file) in a single load operation. However, you need to define a separate load operation for each individual Excel file.

- It is possible to load multiple identically structured Excel files simultaneously. This is explained in Chapter 12.

Note   You cannot load "free-form" Excel worksheets. The data that you load has to be in a tabular format or it will likely be unusable in Power BI.

Microsoft Access Databases

Another widely used data repository that proliferates in many corporations today is Microsoft Access. It is a powerful desktop relational database and can contain hundreds of tables, each containing millions of records. So we need to see how to load data from this particular source. Moreover, Power BI Desktop can be particularly useful when handling Access data because it allows you to see the contents of Access databases without even having to install Access itself.

1. In the Power BI Desktop ribbon, click Get Data ➤ Database and select Access database in the Get Data dialog.

2. Click Connect and navigate to the MS Access database containing the data that you want to load (C:\PowerBIDesktopSamples\ ClientsDatabase.accdb in this example).

3. Select the Access file and click Open. The Navigator dialog appears; it lists all the tables and queries in the Access database.

4. Check the check box for the ClientList table. This displays the contents of the table, as you can see in Figure 2-7.

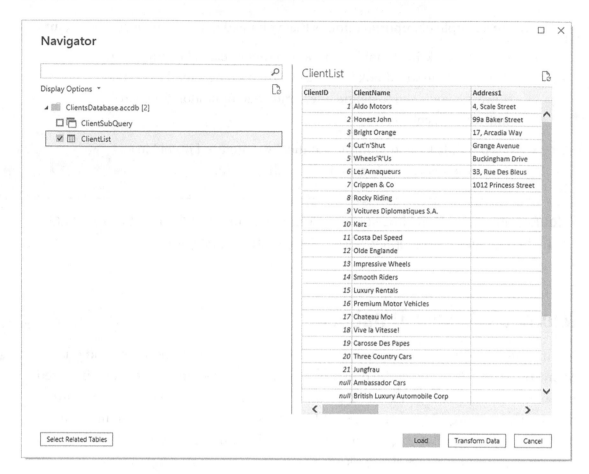

Figure 2-7. *The Navigator dialog before loading data from an Access database*

5. Click Load. The Power BI Desktop window opens and displays the table in the Fields list in the Report window.

If you look closely at the left of the Navigator dialog in Figure 2-7, you can see that it displays two different icons for Access objects:

- A table for Access data tables

- Two small windows for Access queries

This can help you to understand the type of data that you are looking at inside the Access database.

Note   Power BI Desktop *cannot* see linked tables in Access, only imported tables or tables that are actually contained in the Access database. It can, however, read *queries* overlaid upon native, linked, or imported data.

PDF Files

Power BI Desktop can also import *tabular* data from PDF files. This is really amazingly simple to do.

1. In the Power BI Desktop ribbon, click Get Data ➤ File and select PDF in the Get Data dialog.

2. Click Connect and navigate to the PDF document containing the data that you want to load (WordSourceWithTables.Pdf from the sample files in this example).

3. Select the PDF file and click OK. The Navigator dialog appears; it lists all the tables in the PDF document.

4. Check the check box(es) for the table(s) contained in the PDF document that you want to import into Power BI Desktop. This displays the contents of the last selected table, as you can see in Figure 2-8.

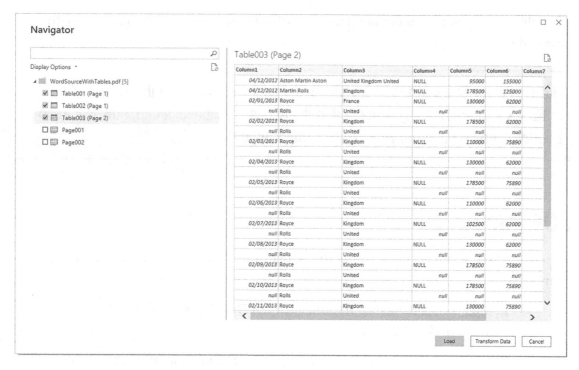

Figure 2-8. *Displaying the tables in a PDF document*

5.   Click Load. The Power BI Desktop window opens and displays
     the table(s)—and their columns as fields—in the Fields list in the
     Report window.

Note   You can only see and consequently import *tables* from PDF documents. You
cannot import paragraphs of text.

I am sure that you can see a pattern emerging in the course of this chapter. Indeed,
this pattern will continue as you progress to loading tables from relational databases in
Chapter 3. The process is nearly always

1.   Know the type of source data that you want to look at

2.   Find the source file that lets you access the data

3.   Examine the data and select the elements that you want to load

Note   You need to be aware that Power BI Desktop can take quite a while to read a large PDF file. So be prepared to be patient when waiting for the Navigator dialog to appear once you have selected the PDF document that you want to load data from.

JSON Files

JSON files are, like XML, a file format that allows users (and computers) to send complex data structures between systems. Generally, JSON files require a little tweaking for them to be loaded in a state that is usable by Power BI Desktop. So we will be looking at how to load and prepare JSON files in Chapter 12, once you have assimilated the necessary data transformation techniques in Chapters 8 through 11.

Conclusion

In this chapter, you have seen how this powerful addition to the Microsoft business intelligence toolset, Power BI Desktop, can help you find and load data from a variety of file-based data sources. These sources can be Access, Excel (including Power View, Power Query, and Power Pivot elements), CSV, XML, PDF, or text files.

You have seen that Power BI Desktop will let you see a sample of the contents of the data sources that it can read without needing any other application. This makes it a superb tool for peeking into data sources and deciding if a file actually contains the data that you need. Indeed, Power BI Desktop's Navigator can help you filter multiple tables in Excel or XML files or Access databases, preview each table, and only select the ones that you want to load. Of course, it can also load dozens of tables at once if they all are stored in the same source.

This chapter is not a complete overview of how to load file-based sources. So if you need to load complex XML files or JSON files or need to understand how to load the contents of entire folders—or all the worksheets in an Excel file, for instance—then you can skip straight to Chapter 12 to learn these techniques.

However, file-based data sources are only a small part of the picture. Power BI Desktop can also load data from a wide range of relational databases and data warehouses. We will take a look at some of these in the next chapter.

CHAPTER 3

Loading Data from Databases and Data Warehouses

Much of the world's corporate data currently resides in relational databases, data warehouses, and data warehouse appliances either on-premises or in the cloud. Power BI Desktop can connect to most of the world's leading databases and data warehouses. Not only that, but it can also connect to many of the lesser-known or more niche data sources that are currently available. This chapter will show you how to extract data from several of these data sources to power your analytics using Power BI Desktop. Indeed, you will discover that once you have learned how to connect to one or two databases, you have learned how to use nearly all of them, thanks to the standardized interface and approach that Power BI Desktop brings to data extraction.

You need to be aware, however, that the examples in this chapter do *not* use sample data that is available on the Apress website. In this chapter, I will let you load your own data or use the sample data that can be installed with the databases themselves.

Note It may be stating the obvious, but connecting to a database means that the database must be installed, set up, and running correctly and you already have to have access to it. Indeed, you may also need specific client software installed on the PC that is running Power BI Desktop. This chapter will *not* explain how to install or use any of the databases (or the client software) that are referenced. For this, you will have to consult the relevant database documentation.

© Adam Aspin 2022
A. Aspin, *Pro Data Mashup for Power BI*, https://doi.org/10.1007/978-1-4842-8578-7_3

Relational Databases

Being able to access the data stored in relational databases is essential for much of today's business intelligence. As enterprise-grade relational databases still hold much of the world's data, you really need to know how to tap into the vast mines of information that they contain. The bad news is that there are many, many databases out there, each with its own intricacies and quirks. The good news is that once you have learned to load data from *one* of them, you can reasonably expect to be able to use *any* of them.

In the real world, connecting to corporate data could require you to have a logon name and usually a password that will let you connect (unless the database can recognize your Windows login or a single sign-on solution has been implemented). I imagine that you will also require permissions to read the tables and views that contain the data. So the techniques described here are probably the easy bit. The hard part is convincing the guardians of corporate data that you actually *need* the data and you should be allowed to see it.

Some of the databases that Power BI Desktop can currently connect to, and can preview and load data from, are given in Table 3-1.

Table 3-1. *Database Sources*

Database	Comments
SQL Server database	Lets you connect to a Microsoft SQL Server on-premises database and import records from all the data tables and views that you are authorized to access.
Access database	Lets you connect to a Microsoft Access file on your network and load queries and tables (which we explored in the previous chapter).
SQL Server Analysis Services database	Lets you connect to a SQL Server Analysis Services (SSAS) data warehouse. This can be either an online analytical processing (OLAP) cube or an in-memory tabular data warehouse.
Oracle database	Lets you connect to an Oracle database and import records from all the data tables and views that you are authorized to access. This will likely require client software to be installed.
IBM DB2 database	Lets you connect to an IBM DB2 database and import records from all the data tables and views that you are authorized to access. This will likely require additional software to be installed.

(*continued*)

Table 3-1. (*continued*)

Database	Comments
IBM Informix database	Lets you connect to an IBM Informix database and import records from all the data tables and views that you are authorized to access. This will likely require additional software to be installed.
IBM Netezza	Lets you connect to an IBM Netezza data warehouse appliance and import records from all the data tables that you are authorized to access. This will likely require additional software to be installed.
MySQL database	Lets you connect to a MySQL database and import records from all the data tables and views that you are authorized to access. This will likely require additional software to be installed.
PostgreSQL database	Lets you connect to a PostgreSQL database and import records from all the data tables and views that you are authorized to access. This will likely require additional software to be installed.
Sybase database	Lets you connect to a Sybase database and import records from all the data tables and views that you are authorized to access. This will likely require additional software to be installed.
Teradata database	Lets you connect to a Teradata database and import records from all the data tables and views that you are authorized to access. This will likely require additional software to be installed.
SAP HANA database	Lets you connect to a SAP HANA in-memory database and import records from all the objects that you have permission to access. This will likely require additional software to be installed.
SAP Business Warehouse Application Server	Allows you to connect to a SAP Business Warehouse. This will likely require client software to be installed.
SAP Business Warehouse Message Server	Allows you to connect to a SAP Business Warehouse. This will likely require client software to be installed.
Amazon Redshift	Lets you connect to an Amazon Redshift database and import records from all the data tables and views that you are authorized to access.

(*continued*)

Table 3-1. (*continued*)

Database	Comments
Impala	Lets you connect to an Impala database and import records from all the data tables and views that you are authorized to access.
Google BigQuery	Lets you access Google BigQuery data repositories.
Essbase	Allows you to connect to Essbase cubes.
Snowflake	Lets you connect to Snowflake data warehouses in the cloud.
Exasol	Lets you connect to Exasol data repositories.

There are many other database connectors that are currently available. As a complete list would take several pages, I advise you to scroll down through the list in the Get Data dialog to see if the database that you are using has a connector to Power BI.

Note   As the list of database and data warehouse sources that you can connect to from Power BI Desktop continues to evolve, this list could likely be extended to include several new items by the time that you read this book.

As well as connections for specific databases, Power BI Desktop contains generic connectors that can help you to read data from databases that are not specifically in the list of available databases. These generic connectors are explained in Table 3-2.

Table 3-2. *Generic Database Access*

Source	Comments
ODBC data source	Lets you connect over Open Database Connectivity to a database or data source
OLE DB data source	Lets you connect over Object Linking and Embedding Database to a database or data source

Be warned that these generic connectors will not work with any database. However, they should work with a database for which you have procured, installed, and configured a valid ODBC or OLE DB driver. You can learn more about this in Chapter 6.

Note   Although Power BI Desktop classifies Microsoft Access as a relational database, I prefer to handle it as a file-based source. For this reason, MS Access data was discussed in the previous chapter.

SQL Server

Here, I will use the Microsoft enterprise relational database—SQL Server—as an example to show you how to load data from a database into Power BI Desktop. The first advantage of this setup is that you probably do not need to install any software to enable access to SQL Server (although this is not always the case, so talk this through with your IT department). The second advantage is that the techniques are pretty similar to those used and applied by Oracle, DB2, and the other databases to which Power BI Desktop can connect. Furthermore, you can load multiple tables or views from a database at once. To see this in action (on your SQL Server database), take the following steps:

1. Open a new Power BI Desktop application.

2. In the Power BI Desktop Home ribbon, click SQL Server. Alternatively, click the small triangle at the bottom of the Get Data button and then click SQL Server. The SQL Server database dialog will appear.

3. Enter the server name in the Server text box. This will be the name of your SQL Server or one of the SQL Server resources used by your organization.

4. Enter the database name. The dialog will look like Figure 3-1 (but with your server and database names, of course).

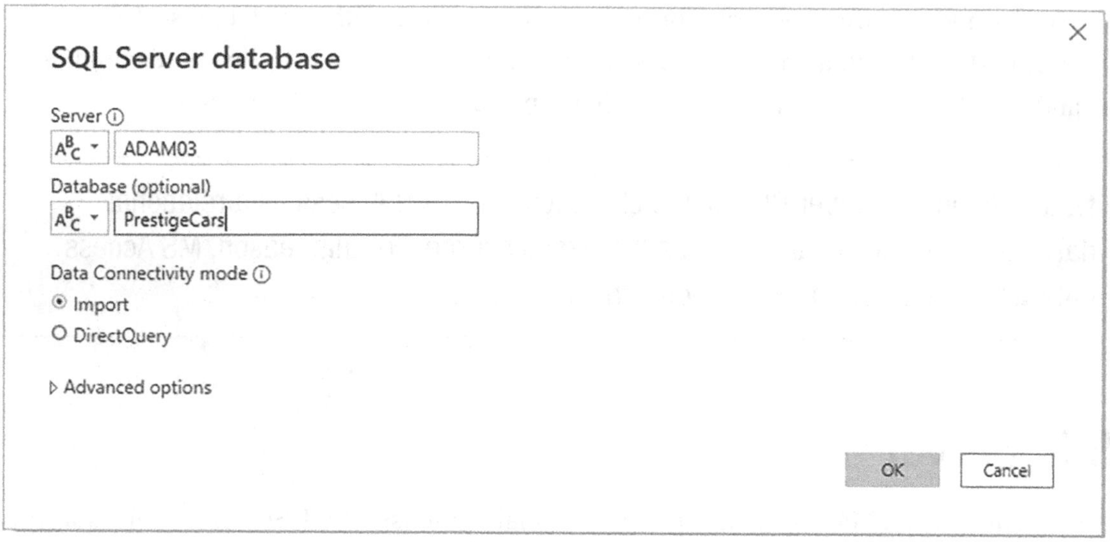

Figure 3-1. *The Microsoft SQL Server database dialog*

5. Click OK. The SQL Server database dialog will appear. Assuming that you are authorized to use your Windows login to connect to the database, leave "Use my current credentials" selected in the left pane, as shown in Figure 3-2.

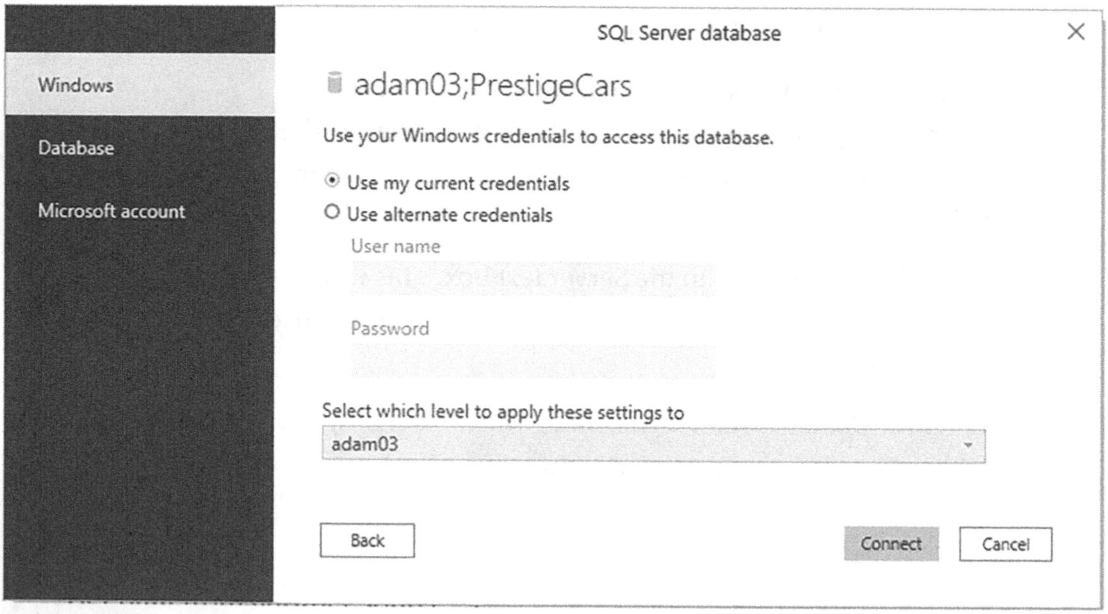

Figure 3-2. *The credentials Database dialog*

6. Click Connect. The Encryption Support dialog may appear to warn you that the connection is not encrypted, as shown in Figure 3-3.

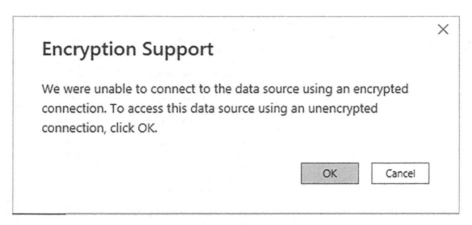

Figure 3-3. *The Encryption Support dialog*

7. Click OK. Power BI Desktop will connect to the server and display the Navigator dialog containing all the tables and views in the database that you have permission to see on the server you selected.

8. Click the check boxes for the tables that you want to load. The data for the most recently selected dataset appears on the right of the Navigator dialog, as shown in Figure 3-4.

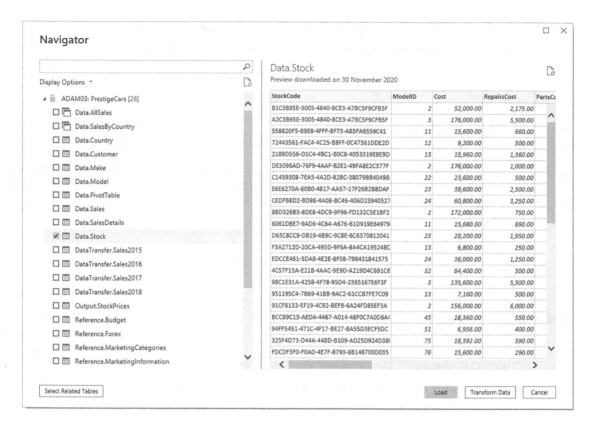

Figure 3-4. *The Navigator dialog when selecting multiple items*

9.   Click Load.

10.   While the data is being loaded, Power BI Desktop will display the
      Load dialog and show the load progress for each selected table.
      You can see this in Figure 3-5.

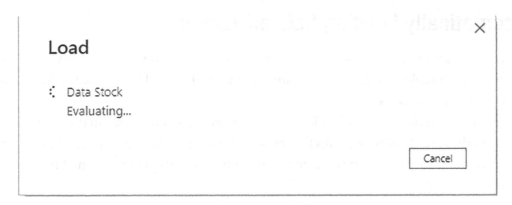

Figure 3-5. *The Load dialog displaying data load progress*

11. The Power BI Desktop window will open and display the tables
     that you selected in the Fields list in the Report window when you
     click OK.

Tip   When selecting multiple tables or views, you will only ever see the contents
of a single data source in the Navigator dialog. However, you can preview the
contents of any of the selected data sources (or even any that are not selected)
simply by clicking the table or view name. This will not affect the choice of *selected*
tables and views that you want to load into Power BI Desktop.

Since this is very similar to the way in which you loaded data from Access in the
previous chapter, I imagine that you are getting the hang of how to use database sources
by now. Once again, the Navigator dialog is a simple and efficient way to select the
datasets that you want to use in your reports and dashboards.

Note   You can enter the server IP address instead of the server name if you
prefer. If there are several SQL Server instances on the same server, you will need
to add a backslash and the instance name. This kind of detailed information can be
obtained from corporate database administrators (DBAs).

Automatically Loading Related Tables

Relational databases are nearly always intricate structures composed of many interdependent tables. Indeed, you will frequently need to load several tables to obtain all the data that you need.

Knowing which tables to select is not always easy. Power BI Desktop tries to help you by automatically detecting the links that exist in the source database between tables; this way, you can rapidly isolate the collections of tables that have been designed to work together.

Do the following to see a related group of tables:

1. Connect to the source database as described in the previous section.

2. In the Navigator dialog, click a table that contains data that you need.

3. Click the "Select related tables" button.

Any tables in the database that are linked to the tables that you selected in the Navigator dialog are selected. You can deselect any tables that you do not want, of course. More importantly, you can click the names of the selected tables to see their contents.

Note   Sometimes, you have to select several tables in turn and click "Select related tables" to ensure that Power BI Desktop will select all the tables that are necessary to underpin your analysis.

Database Options

The world of relational databases is—fortunately or unfortunately—a little more complex than the world of text and CSV files or MS Access data sources. Consequently, there are a few comments to make about using databases as a data source—specifically, how to connect to them.

First, let's cover the initial connection to the server. The options are explained in Table 3-3.

Table 3-3. *Database Connection Options*

Option	Comments
Server	You cannot browse to find the server, and you need to type or paste the server name. If the server has an instance name, you need to enter the server and the instance. Your IT department will be able to supply this if you are working in a corporate environment.
Database	If you know the database, then you can enter (or paste) it here. This restricts the number of available tables in the Navigator dialog and makes finding the correct table or view easier. Otherwise, all the accessible databases—and all the accessible tables they contain—will be visible.
SQL statement	You can enter a valid snippet of T-SQL (or a stored procedure or a table-valued function) that returns data from the database.

These options probably require a little more explanation. So let's look at each one in turn.

Server Connection

It is fundamental that you know the exact connection string for the database that you want to connect to. This could be the following:

- The database server name.

- The database server name, a backslash, and an instance name (if there is one).

- The database server IP address.

- The database server IP address, a backslash, and an instance name (if there is one).

- If the SQL Server instance is using a custom port, you must end the server name with a comma followed by the port number. This is, inevitably, a question for corporate DBAs.

- If you are running a single SQL Server instance on your own PC, then you can use the name *localhost* (or a period) to refer to the local server.

Note   A database instance is a separate SQL Server service running alongside others on the same physical or virtual server. You will always need both the server and this instance name (if there is one) to successfully connect. You can also specify a timeout period if you wish.

Most SQL Server instances host many, many databases. Sometimes, these can number in the hundreds. Sometimes, inevitably, you cannot remember which database you want to connect to. Fortunately, Power BI Desktop can let you browse the databases on a server that you are authorized to access. To do this, do the following:

1.   In the Power BI Desktop ribbon, click the small triangle at the bottom of the Get Data button and then click SQL Server (or just click the SQL Server button in the Home ribbon). The SQL Server database dialog will appear.

2.   Enter the server name in the Server text box and click OK. Do *not* enter a database name. The Navigator window opens and displays all the available databases, as shown in Figure 3-6. Of course, the actual contents depend on the server that you are connecting to.

Figure 3-6. *The Navigator dialog when selecting databases*

You can see from Figure 3-6 that if you click the small triangle to the left of a database, then you are able to see all the tables and views that are accessible to you in this database. Although this can mean an overabundance of possible choices when looking for the table(s) or view(s) that you want, it is nonetheless a convenient way of reminding you of the name of the dataset that you require.

Tip   The actual databases that you will be able to see on a corporate server will depend on the permissions that you have been given. If you cannot see a database, then you will have to talk to the database administrators to sort out any permission issues.

Searching for Databases, Tables, and Views in Navigator

If you are overwhelmed by the sheer volume of tables and views that appear in the left panel of the Navigator dialog, then you can use Navigator's built-in search facility to help you to narrow down the set of potential data sources.

Searching for Databases

To isolate specific databases, do the following:

1. Carry out steps 1 and 2 in the earlier "SQL Server" section to connect to a SQL Server instance *without* specifying a database.

2. In the Search box of the Navigator dialog, enter a few characters that you know are contained in the name of the database that you are looking for. Entering, for example, **Prest** on my server gives the result that you see in Figure 3-7.

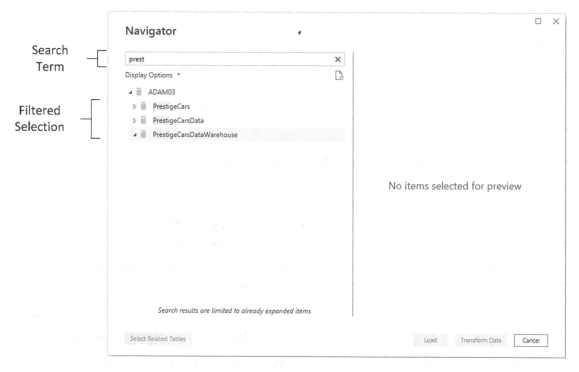

Figure 3-7.   *Using Search with Navigator to find databases*

Searching for Tables

If you are searching for tables, do the following:

1. In the Search box of the Navigator dialog, enter a few characters that you know are contained in the name of the table or view that you are looking for. Entering, for example, **make** on my server gives the result that you see in Figure 3-8.

Filtered Tables and Views

Figure 3-8. *Using Search with Navigator to find tables*

When searching for objects, you can enter the text in uppercase or lowercase (at least, this is the case with most SQL Server implementations), and the text can appear anywhere in the names of the tables or views—not just at the start of the name. With every character that you type, the list of potential matches gets shorter and shorter. Once you have found the table or view that you are looking for, simply proceed as described earlier to load the data into Power BI Desktop.

If your search does not return the subset of tables in any views that you were expecting, all you have to do is click the cross at the right of the Search box. This cancels the search and displays all the available tables, as well as clears the Search box.

If you are not convinced that you are seeing all the tables and views that are in the database, then click the small icon at the bottom right of the Search box (it looks like a small page with two green circular arrows). This is the Refresh button, which refreshes the connection to the database and displays all the tables and views that you have permission to see. Finally, it is worth noting that filtering tables can also be applied to most data sources and not just databases. This covers Excel tables, worksheets, and named ranges–among many other data sources. This is another example of how the unified Navigator interface can help minimize the learning curve when it comes to mastering Power BI Desktop.

Note   SQL Server databases can also be accessed using the DirectQuery option. This technique is explained in the next chapter.

Database Security

Remember that databases are designed to be extremely secure. Consequently, you only see servers, databases, tables, and views if you are authorized to access them. You might have to talk to your IT department to ensure that you have the required permissions; otherwise, the table that you are looking for could be in the database but remain invisible to you.

Tip   If you experience a connection error when first attempting to connect to SQL Server, simply click the Edit button to return to the Microsoft SQL Database dialog and correct any mistakes. This avoids having to start over.

Using a SQL Statement

If there is a downside to using a relational database such as SQL Server as a data source, it is that the sheer amount of data that the database stores—even in a single table—can be dauntingly huge. Fortunately, all the resources of SQL Server can be used to filter the data that is used by Power BI Desktop before you even load the data. This way, you do not have to load entire tables of data at the risk of drowning in information before you have even started to analyze it.

The following are SQL Server techniques that you can use to extend the partnership between SQL Server and Power BI Desktop:

- SQL SELECT statements

- Stored procedures

- Table-valued functions

These are, admittedly, fairly technical solutions. Indeed, if you are not a database specialist, you could well require the services of your IT department to use these options to access data in the server. Nonetheless, it is worth taking a quick look at these techniques in case they are useful now or in the future.

Any of these options can be applied from the SQL Server database dialog. Here is an example of how to filter data from a database table using a SELECT statement:

1. In the Power BI Desktop ribbon, click the small triangle at the bottom of the Get Data button and then click SQL Server. The SQL Server database dialog will appear.

2. Enter the server name and the database.

3. Click the triangle to the left of Advanced options. This opens a box where you can enter a SQL command.

4. Enter the SQL command that you want to apply. In this case, it is SELECT CountryName, MakeName, ModelName, Cost FROM Data.AllSales ORDER BY CountryName. This custom SQL will, of course, only work on my database. You will need to enter your own custom SQL SELECT clause. The dialog will look like Figure 3-9.

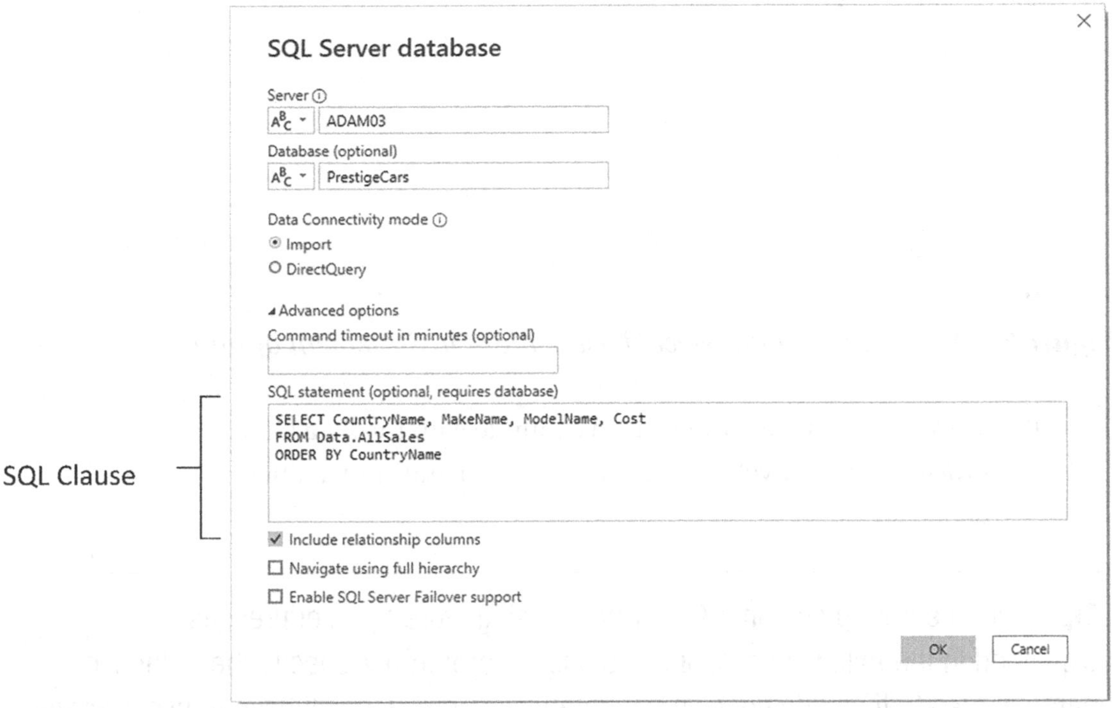

Figure 3-9. Using SQL to select database data

5.   Click OK. You may get the Credentials and Encryption dialogs appearing. A sample of the corresponding data is eventually displayed in a dialog like the one shown in Figure 3-10.

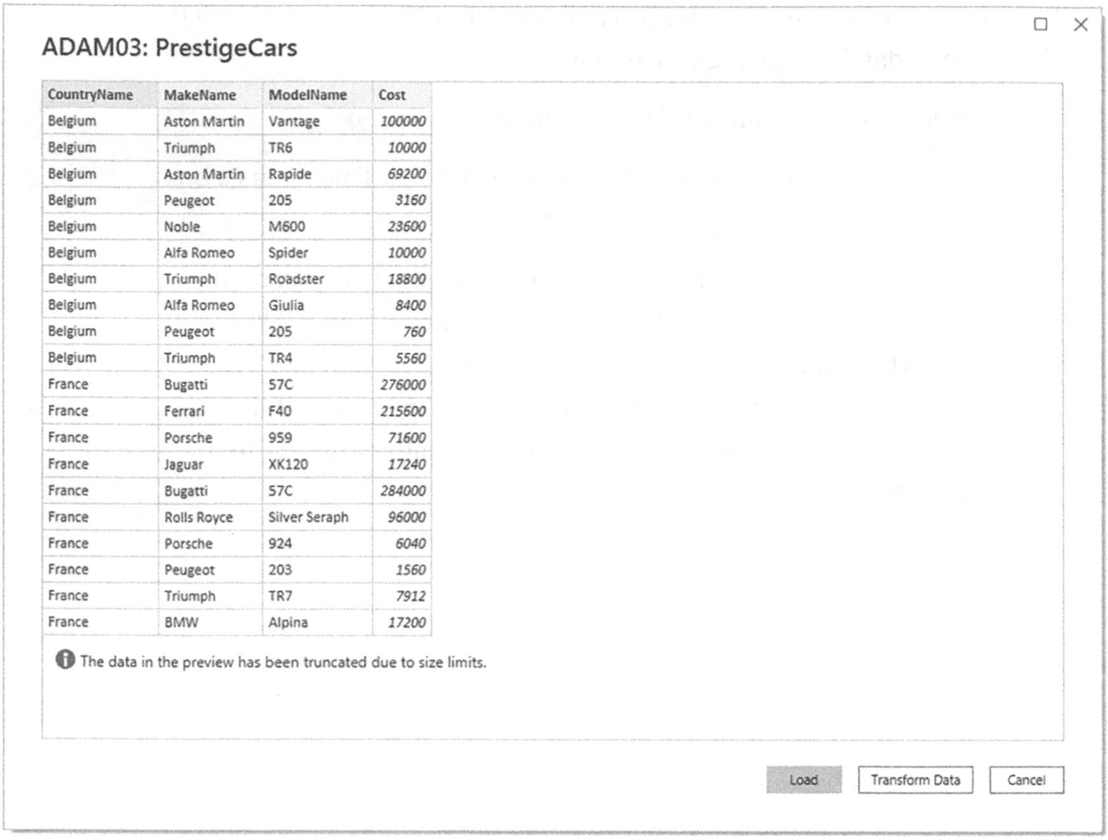

ADAM03: PrestigeCars

CountryName	MakeName	ModelName	Cost
Belgium	Aston Martin	Vantage	100000
Belgium	Triumph	TR6	10000
Belgium	Aston Martin	Rapide	69200
Belgium	Peugeot	205	3160
Belgium	Noble	M600	23600
Belgium	Alfa Romeo	Spider	10000
Belgium	Triumph	Roadster	18800
Belgium	Alfa Romeo	Giulia	8400
Belgium	Peugeot	205	760
Belgium	Triumph	TR4	5560
France	Bugatti	57C	276000
France	Ferrari	F40	215600
France	Porsche	959	71600
France	Jaguar	XK120	17240
France	Bugatti	57C	284000
France	Rolls Royce	Silver Seraph	96000
France	Porsche	924	6040
France	Peugeot	203	1560
France	Triumph	TR7	7912
France	BMW	Alpina	17200

ⓘ The data in the preview has been truncated due to size limits.

Load   Transform Data   Cancel

Figure 3-10. *Database data selected using the SQL Statement option*

6.   Click Load or Transform Data to continue with the data load process. Alternatively, you can click Cancel and start a different data load.

Tip   When entering custom SQL (or when using stored procedures, as is explained in the following section), you should, preferably, specify the database name in step 3. If you do not give the database name, you will have to use a three-part notation in your SQL query. That is, you must add the database name and a period before the schema and table name of *every* table name used in the query.

Stored Procedures in SQL Server

The same principles apply when using stored procedures of functions to return data from SQL Server. You will always use the SQL Statement option to enter the command that will return the data. Just remember that to call a SQL Server stored procedure or function, you would enter the following elements into the Microsoft SQL Database dialog:

- *Server*: <your server name>

- *Database*: <the database name>

- *SQL Statement*: EXECUTE (or EXEC) <enter the schema (if there is one, followed by a period) and the stored procedure name, followed by any parameters>

This way, either you or your IT department can create complex and secure ways to allow data from the corporate databases to be read into Power BI Desktop from databases.

To see this in practice, here is an example of using a SQL Server stored procedure to return only a subset of the available data. The stored procedure is called pr_DisplayUKClientData, and you apply it like this:

1. In the Power BI Desktop ribbon, click the small triangle at the bottom of the Get Data button and then click SQL Server. The SQL Server database dialog will appear.

2. Enter the server name and the database.

3. Click the triangle to the left of Advanced options. This opens a box where you can enter a SQL command.

4. Enter the SQL command that you want to apply. In this case, it is EXECUTE `dbo.pr_DisplayUKClientData`. The dialog will look like Figure 3-11.

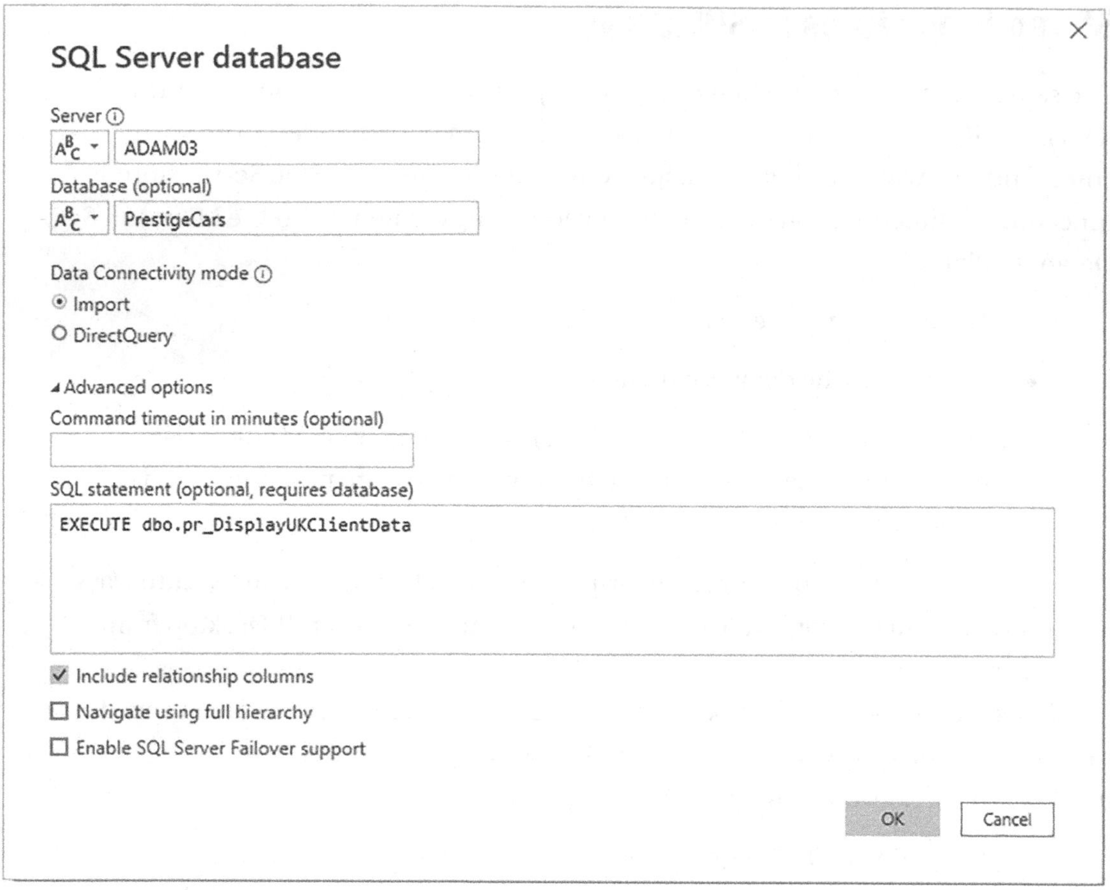

Figure 3-11. *Using SQL to select database data using a stored procedure*

5. Click OK. A sample of the corresponding data is returned to the Navigator.

6. Click Load or Transform Data to continue with the data load process. Alternatively, you can click Cancel and start a different data load.

The data that is returned in this example is only a subset of the available data that has been selected by the stored procedure. You need to be aware that stored procedures can perform a multitude of tasks on the source data. These can include selecting, sorting, and cleansing the data.

Note   You will need to use stored procedures from your SQL Server database to return data. Stored procedures can be created and supplied by your IT department.

Stored procedures often require *parameters* to be added after the stored procedure name. This is perfectly acceptable when executing a stored procedure in Power BI. An example of one way of using a parameter in SQL Server would be

```
EXECUTE dbo.pr_DisplayUKClientData 2020
```

The key thing to remember—and to convey to your IT department—is that the SQL that Power BI expects is the flavor of SQL that the source database uses. So for SQL Server, that means using T-SQL. In fact, this SQL becomes a "pass-through" query that is interpreted directly by the underlying database.

Note   A SQL statement or stored procedure will only return data as a *single table*. Admittedly, this table could contain data from several underlying tables or views in the source database, but filtering the source data will prevent Power BI Desktop from loading data from several tables as separate queries. Consequently, you could have to create multiple queries rather than a single load query to get data from a coherent set of tables in the data source as Power BI will only accept the first result set from a stored procedure (if the stored procedure returns multiple tables of data).

Oracle Databases

There are many, many database vendors active in the corporate marketplace today. Arguably the most dominant of them is currently Oracle. While I have used Microsoft data sources to begin the journey into an understanding of how to use databases with Power BI Desktop, it would be remiss of me not to explain how to access databases from other suppliers.

So now is the time to show you just how open-minded Power BI Desktop really is. It does not limit you to Microsoft data sources—far from it. Indeed, it is every bit as easy to use databases from other vendors as the source of your analytical reports. As an example of this, let's take a look at loading Oracle data into Power BI Desktop.

Installing and configuring an Oracle database is a nontrivial task. Consequently, I am not providing an Oracle sample database but will leave you either to discover a corporate database that you can connect to or, preferably, consult the many excellent resources available that do an excellent job of explaining how to set up your own Oracle database and install the sample data that is available.

Be aware that connecting to Oracle will require installing Oracle client software on the computer where you are running Power BI Desktop. This, too, can be complex to set up. So you might need some help from a corporate resource if you are planning to use Oracle data with Power BI Desktop.

Should you be feeling brave, you can use the following URLs to find the Oracle client software. For 64-bit versions of Power BI Desktop, use the following link to download and install the 64-bit Oracle client:

`www.oracle.com/database/technologies/odac-downloads.html`

This link was active as this book went to press.

If you need to check which version of Power BI Desktop you are using (32 bit or 64 bit), click File ➤ Help ➤ About. You will see a dialog that tells you which version you are using.

So assuming that you have an Oracle database available (and that you know the server name or SID as well as a valid username and password), the following steps show how you can load data from this particular source into Power BI Desktop. I will be using standard Oracle sample data that is often installed with sample databases in this example.

1. Open a new Power BI Desktop application.

2. In the Power BI Desktop Home ribbon, click Get Data.

3. In the Get Data dialog, click Database on the left.

4. Click Oracle database on the right. The dialog will look like Figure 3-12.

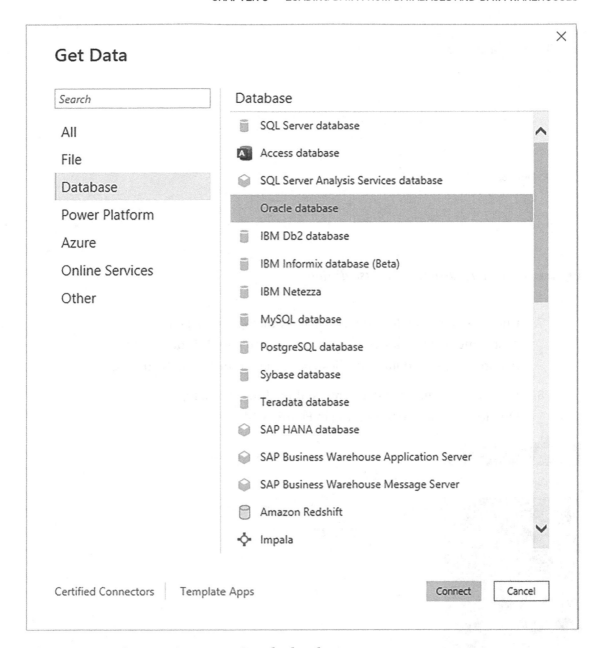

Figure 3-12. *Connecting to an Oracle database*

5. Click Connect. The Oracle database dialog will appear.

6. Enter the server name in the Server text box. This will be the name of your Oracle server or one of the Oracle server resources used by your organization. The dialog will look like Figure 3-13.

Figure 3-13. *The Oracle database dialog*

7. Click OK. The Oracle database security dialog will appear. Assuming that you are not authorized to use your Windows login to connect to the database, click Database on the left of the dialog.

8. Enter the username and password that allow you to log in to Oracle. You can see this dialog in Figure 3-14.

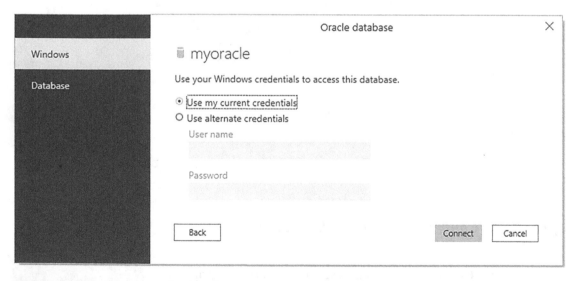

Figure 3-14. *The Oracle database security dialog*

9. Click Connect. Power BI Desktop will connect to the server and display the Navigator dialog containing all the tables and views in the database that you have permission to see on the server you selected. In some cases, you could see a dialog saying that the data source does not support encryption. If you feel happy with an unencrypted connection, then click the OK button for this dialog.

10. Expand the HR folder. This is a standard Oracle sample schema that could be installed on your Oracle instance. If not, you will have to choose another schema. Click the check boxes for the tables that interest you. The data for the most recently selected data appears on the right of the Navigator dialog, as shown in Figure 3-15.

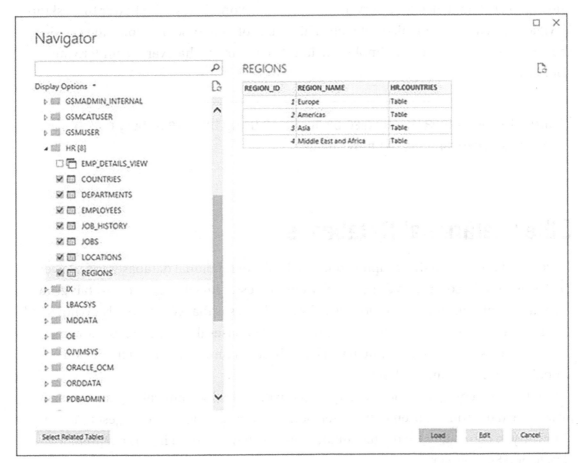

Figure 3-15. *The Navigator dialog using Oracle data*

11. Click Load. The Power BI Desktop window will open and display the tables that you selected in the Fields list in the Report window when you click OK.

If you have already followed the example earlier in this chapter to load data from SQL Server, you will probably appreciate how much the two techniques have in common. Indeed, one of the great advantages of using Power BI Desktop is that loading data from different data sources follows a largely similar approach and uses many of the same steps and dialogs. This is especially true of databases, where the steps are virtually identical—whatever the database.

Of course, no two relational databases are completely alike. Consequently, you connect to an Oracle instance (or server) but cannot choose a database as you can in SQL Server (or Sybase, for instance). Similarly, where Oracle has schemas to segregate and organize data tables, SQL Server has databases. Nonetheless, the Power BI Desktop Navigator will always organize data into a hierarchy of folders so that you can visualize the data structures in a clear, simple, and intuitive manner, whatever the underlying database.

Note   Oracle databases can also be accessed using the DirectQuery option. This technique is explained in the next chapter.

Other Relational Databases

Table 3-1 at the start of this chapter contains the list of relational databases that Power BI Desktop could connect to as this book went to press. I imagine that the list has grown since this book was published. However, the good news is that you probably do not need much more information to connect to any of the databases that are available for you to use as data sources. Simply put, if you know how to connect to one of them, you can probably connect to any of them.

So I am not going to fill out reams of pages with virtually identical explanations of how to get data from a dozen or more relational databases. Instead, I suggest that you simply try to connect using the techniques that you have learned in this chapter for Oracle and SQL Server.

Be warned, though, that to connect to a relational database, you will inevitably need to know the following details:

- The server name

- A database name (possibly)

- A valid username (depending on the security that has been implemented)

- A valid password for the user that you are connecting as (this, too, will depend on the security in place)

However, if you have these elements, then nothing should stop you from using a range of corporate data sources as the basis for your analysis with Power BI Desktop. You will, of course, need all the necessary permissions to access the database and the data that it contains.

It is also worth knowing that connecting to DB2, MySQL, PostgreSQL, Sybase, IBM Informix, IBM Netezza, SAP HANA, or Teradata can require not only that the database administrator has given you the necessary permissions but also that connection software (known as *drivers* or *providers*) has been installed on your PC. Given the "corporate" nature of the requirements, it may help if you talk directly to your IT department to get this set up in your enterprise IT landscape.

One way to find out if the software that is required to enable a connection to a specific database has been installed is to select the database from the list available in the Get Data dialog. If the drivers have not been installed, you will see a warning similar to the one in Figure 3-16.

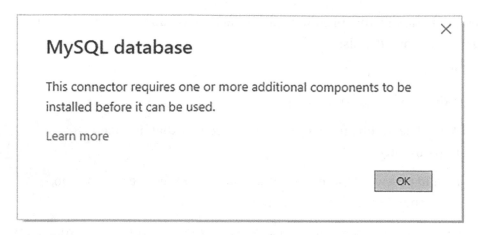

Figure 3-16. *The missing driver alert*

Clicking the "Learn more" link will take you to the download page for the missing drivers. You will have to install all the required additional components before attempting to connect to the database again. Be warned, however, that configuring data providers can, in some cases, require specialist knowledge as well as access rights on the computer where the drivers have to be installed.

In some cases, you may see the alert shown in Figure 3-17.

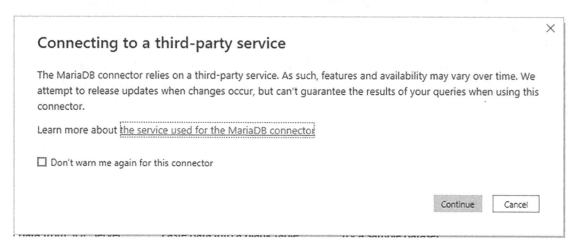

Figure 3-17. *The third-party driver alert*

If you feel happy continuing with the potential limits of such a connection, then continue to load your data.

Microsoft SQL Server Analysis Services Data Sources

An Analysis Services database is a data warehouse technology that can contain vast amounts of data that has been optimized to enable decision-making. *SSAS cubes* (as these databases are also called) are composed of facts (measures or values) and dimensions (descriptive attributes).

In fact—and with apologies to data warehouse purists—an SSAS cube is, essentially, a gigantic pivot table. So if you have used pivot tables in Excel, you are ready to use data warehouse sources in Power BI Desktop.

Note   In this section, I will be explaining access to *dimensional* (disk-based) SSAS data warehouses. I explain *tabular* SSAS in the next section.

If your workspace uses Analysis Services databases, you can access them by doing the following steps:

1. In the Power BI Desktop ribbon, click Get Data ➤ More. Click Database on the left of the Get Data dialog and select SQL Server Analysis Services database in the Get Data dialog.

2. Click Connect. The SQL Server Analysis Services database dialog will appear.

3. Enter the Analysis Services server name and the database (or "cube") name. The database I am using here is called CarSalesOLAP; you will have to specify your own SSAS database name. In any case, you will need to know and use the name of your own SSAS server and database. The dialog will look something like the one shown Figure 3-18.

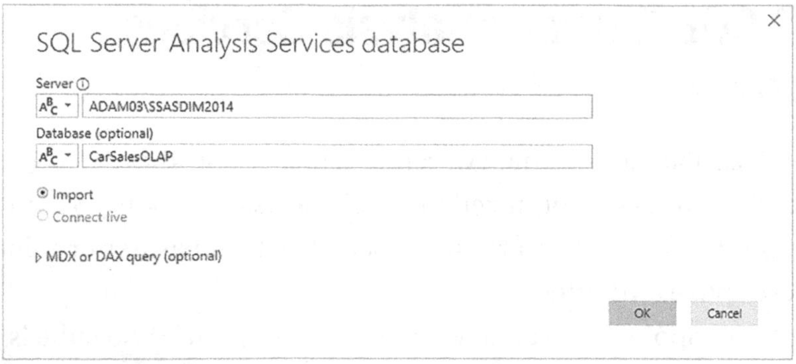

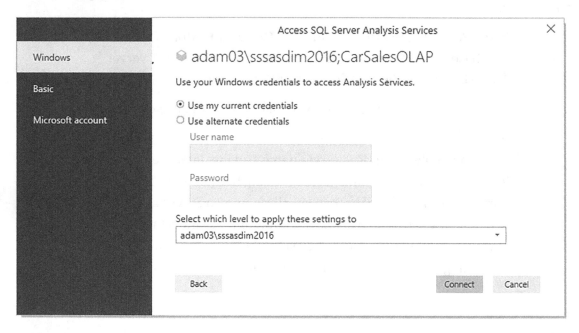

Figure 3-18. *Connecting to an SSAS (multidimensional) database*

4. Click OK. If this is the first time that you are connecting to the cube, then the Access SQL Server Analysis Services dialog will appear so that you can define the credentials that you are using to connect to the Analysis Services database, as shown in Figure 3-19.

Figure 3-19. *SQL Server Analysis Services credentials dialog*

5. Accept or alter the credentials and click Connect. The Navigator dialog will appear.

6. Expand the folders in the left pane of the dialog. This way, you can see all the fact tables and dimensions contained in the data warehouse.

7. Select the fact tables, dimensions, or even only the dimension elements and measures that you want to load. The dialog will look something like Figure 3-20.

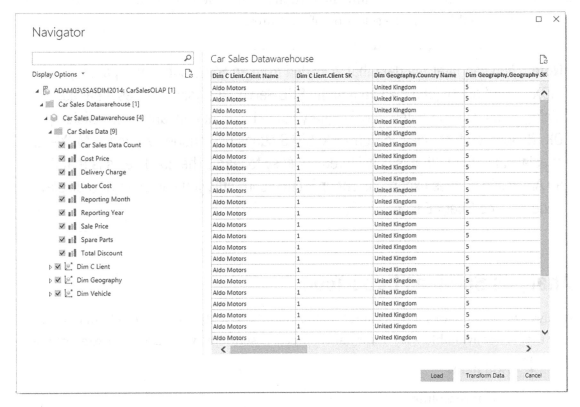

Figure 3-20. *Selecting attributes and measures from an SSAS cube*

8. Click Load. The Power BI Desktop window will open and display the measures and attributes that you selected in the Fields list in the Report window.

Note   If you did not enter the cube (database) name in step 3, then the Navigator dialog will display all the available cubes on the SSAS server. From here, you can drill down into the cube that interests you to query the data you require.

SSAS cubes are potentially huge. They can contain dozens of dimensions, many fact tables, and literally thousands of measures and attributes. Understanding multidimensional cubes and how they work is beyond the scope of this book. Nonetheless, it is important to understand that for Power BI Desktop, a cube is just another data source. This means that you can be extremely selective as to the cube elements that you load into Power BI Desktop and only load the elements that you need for your analysis. You can load entire dimensions or just a few attributes, just like you can load whole fact tables or just a selection of measures.

Note   You can filter the data that is loaded from an SSAS cube by expanding the MDX or DAX query (optional) item in the SQL Server Analysis Services database dialog. Then you can enter an MDX query in the box that appears before clicking OK. Be warned that "classic" (on-disk) SSAS cubes use queries written in MDX—a specialist language that is considered not easy to learn. The good news is that if an Analysis Services expert has set up a cube correctly, you can see SSAS display folders in the Query Editor.

Analysis Services Cube Tools

Analysis Services data sources allow you to tweak the selection of source elements in a way that is not available with other data sources. Essentially, you have two extra options:

- Add Items
- Collapse Columns

Add Items

When using an SSAS data source, you can at any time add any attributes or measures that you either forgot or thought that you would not need when setting up the initial connection.

1. In Power BI Desktop, click the Edit Queries button. The Power BI
    Desktop Query Editor window will be displayed. Assuming that
    there is only one query, the Manage ribbon will appear, as shown
    in Figure 3-21. Otherwise, click the SSAS query that you have
    previously established.

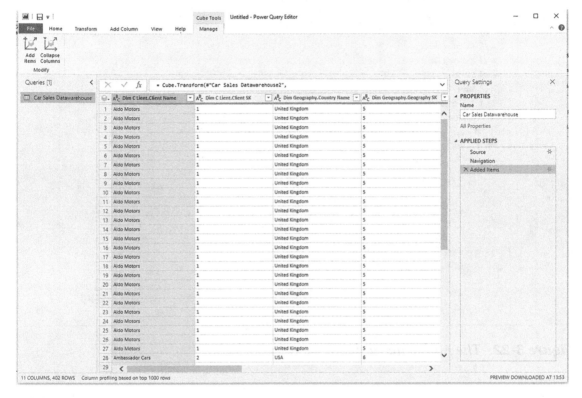

Figure 3-21. *Cube Tools*

2. In the Manage ribbon, click the Add Items button. The Add Items
    dialog will appear.

3. Expand any measure groups and select all the measures and
    attributes that you want to add, as shown in Figure 3-22.

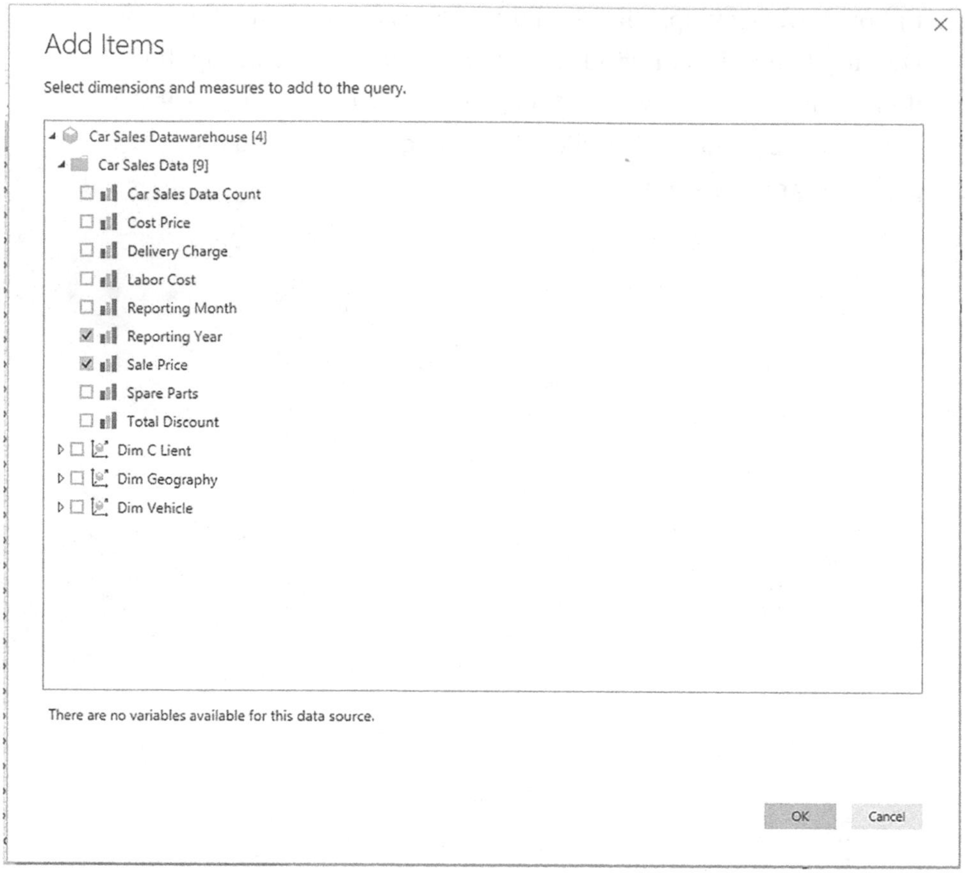

Figure 3-22. *The Add Items dialog*

    4.   Click OK.

    5.   In the Power BI Desktop Query Editor, activate the Home ribbon and click Close & Apply.

Any changes that you made are reflected in the data, and the selected measures and attributes are added as new columns at the right of the dataset.

Note   Power BI Desktop Query will not detect if any new measures and attributes that you add are already in the dataset. So if you add an element a second time, it will appear *twice* in the query.

Collapse Columns

Do the following to remove any columns that you no longer require from the data source (which can accelerate data refresh):

1. In the Query pane, click the SSAS query that you have previously established. (The sample data for this connection will be displayed in the center of the Query window—this is yet another term for the Query Editor window.) The Manage ribbon will appear.

2. In the Manage ribbon, click the Collapse Columns button.

The columns are removed from the connection to the SSAS cube and, consequently, from the Fields list at the right of the Power BI Desktop window. They are also removed from any visualizations that use them.

Note   Removing columns from Power BI Desktop Query can have a serious domino effect on reports and dashboards that prevents visuals from displaying correctly. Consequently, you need to be very careful when removing them.

SSAS Tabular Data Warehouses

The previous section showed you how to connect to a "classic" SQL Server Analysis Services cube. However, there are now two types of SQL Server Analysis Services data warehouses:

- The "traditional" dimensional cube

- The "newer" tabular data warehouse

As more and more data warehouses (at least the ones that are based on Microsoft technologies) are being built using the newer, tabular technology, it is probably worth your while to see how quickly and easily you can use these data sources with Power BI Desktop. Indeed, the steps that you follow to connect to either of these data warehouse sources are virtually identical. However, as Power BI is rapidly becoming the tool of choice to query tabular data warehouses, it is certainly worth a few minutes to learn how to connect to SSAS tabular (as it is often called, for short).

1. In the Power BI Desktop ribbon, click Get Data ➤ More ➤
    Database and select SQL Server Analysis Services database in the
    Get Data dialog.

2. Click Connect. The SQL Server Analysis Services database dialog
    will appear.

3. Enter the Analysis Services server name and the tabular database
    name (we don't tend to call these cubes). Here, the database is
    CarSalesTabular on my PC; you will have to specify your own
    tabular database name. In any case, you will need to use the name
    of your own SSAS server.

4. Click Import.

5. The dialog will look like Figure 3-23.

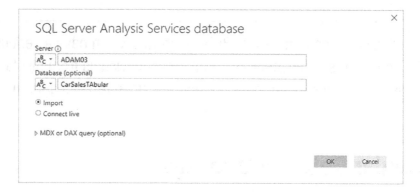

Figure 3-23. *Connecting to an SSAS (multidimensional) database*

6. Click OK. If this is the first time that you are connecting to the
    tabular warehouse, then the Access SQL Server Analysis Services
    dialog will appear so that you can define the credentials that you
    are using to connect to the Analysis Services database, where you
    will have to accept or alter the credentials and click Connect. The
    Navigator dialog will appear.

7. Expand the folders in the left pane of the dialog. This way, you can
    see all the tables contained in the data warehouse. These may—or
    may not—be structured as facts and dimensions as was the case
    with a "classic" SSAS data warehouse.

8. Select the tables that you want to load. The dialog will look something like Figure 3-24.

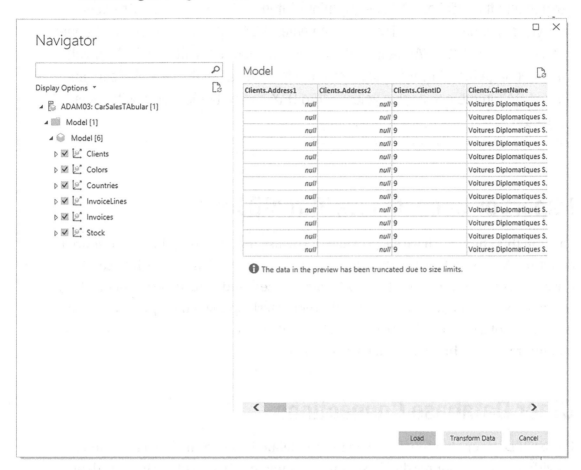

Figure 3-24. *Selecting attributes and measures from an SSAS tabular data source*

9. Click Load. The Power BI Desktop window will open and display the measures and attributes that you selected in the Fields list in the Report window.

Tip   You can filter the data that is loaded from an SSAS tabular database by expanding the MDX or DAX query (optional) item in the SQL Server Analysis Services database dialog. Then you can enter a DAX query in the box that appears before clicking OK. SSAS tabular databases use queries written in DAX, which is the language that Power BI itself uses to filter and calculate data. You can learn about DAX in the companion volume "Pro DAX and Data Modeling with Power BI Desktop."

Import or Connect Live/DirectQuery

So far in this chapter, I have suggested that you use the Import option when sourcing data from Microsoft SQL Server databases and data warehouses. This is because the alternative, Connect Live (for SSAS data sources) and DirectQuery (for database sources), is such an important part of Power BI Desktop that I have preferred to make it the subject of a whole separate chapter. You will discover how to use this far-reaching and impressive technology in the next chapter.

Other Database Connections

Power BI Desktop does not limit you to a predefined set of available data sources. Provided that your source database comes complete with one of the generic data providers—ODBC or OLE DB—then Power BI Desktop can, in all probability, access these sources too. You will learn about these in Chapter 6.

Conclusion

In this chapter, you have seen how to connect Power BI Desktop to some of the plethora of databases and data warehouses that currently exist. Moreover, you have seen that Power BI Desktop comes equipped "out of the box" with connections to most of the databases that currently exist in a corporate environment.

Despite their usefulness in storing and structuring large quantities of information, corporate databases can present one small drawback: the time that it can take to load the data from the database into Power BI Desktop's in-memory data model. The development team at Microsoft is clearly aware of this potential shortcoming and has come up with a solution, called DirectQuery (or Connect Live), which you can discover how to implement in the next chapter.

CHAPTER 4

DirectQuery and Connect Live

The previous chapter showed you how to access data from a range of database and data warehouse sources. This process is both simple and efficient, as you saw. However, there is one stage in the process of fetching data that can take a little time, especially if you are dealing with large datasets. This is the "load" phase where the data from the source system is transferred into the Power BI Desktop in-memory model and compressed.

The Power BI development team has clearly looked hard at this question and has come up with a potentially far-reaching solution: connect directly to the data source and avoid having to download the data. This technique is called DirectQuery. It is worth noting that Microsoft now calls the direct data connection to SQL Server Analysis Services "Connect Live." However, I will consider this to be, nevertheless, part of the DirectQuery technological approach.

In this chapter, we will take a look at when you can use DirectQuery in Power BI Desktop and what the advantages (and, of course, any drawbacks) are to using these data connection methods. Then you will see how to use database sources for Power BI by learning how to swap database connections and modify access permissions when connecting to databases.

DirectQuery and Connect Live

To begin with, you need to know that DirectQuery (and Connect Live) currently only works with a few of the available data sources that Power BI Desktop can connect to. At the time of writing, some of these (and this list is not exhaustive) are

- SQL Server database (on-premises and in Azure)
- SQL Server Analysis Services ("classic" SSAS and tabular)

© Adam Aspin 2022
A. Aspin, *Pro Data Mashup for Power BI*, https://doi.org/10.1007/978-1-4842-8578-7_4

- Power BI Datasets

- Oracle database

- Teradata database

- PostgreSQL

- SAP HANA

- SAP Business Warehouse Message Server

- SAP Business Warehouse Server

- Azure SQL Database

- Azure SQL Data Warehouse (now called Azure Synapse Analytics)

- Amazon Redshift

- AtScale cubes

- Azure HDInsight Spark

- Denodo

- Exasol

- Essbase

- HDInsight Interactive Query

- Impala

- Snowflake

- Spark

- Azure Databricks

- Actian

- Amazon Athena

- Google BigQuery

- IBM Netezza

DirectQuery is different from the more traditional data load methods for the following reasons:

- You do not load the data into Power BI Desktop. Instead, you use the data directly from the database server.

- Because you are not loading a copy of the data into Power BI Desktop, you *cannot* work offline. You need to be able to connect to the source database or data warehouse to use the data.

- The connection to the source database or data warehouse—and the consequent availability of the data for analysis—is usually extremely fast, if not instantaneous.

- The data is fetched specifically for the requirements of each new visual that you create.

- You have all the data that is in the source database or data warehouse available.

- Data is refreshed every time you apply a slicer or a filter.

- You can also connect to other data sources if you are using DirectQuery. This means that you can mix data from DirectQuery sources and those loaded into Power BI Desktop.

There are several very valid reasons why you might want to use DirectQuery:

- It lets you build visualizations over *very large datasets*, where it would otherwise be unfeasible to first import all of the data.

- It allows for enhanced security as data only transits through the Power BI Service and is not duplicated outside the enterprise or its other cloud repositories.

- You are sure of seeing the latest data always.

- It enables near real-time reporting (assuming a fast network connection to a sufficiently powerful database server).

- Security rules are defined in the underlying source that connects to the data source using the current user's credentials or the defined user in the gateway (for many data sources).

- Data sovereignty requirements can be respected.

- It applies data warehouse (SSAS/SAP) measures at the correct granularity.

There is, unfortunately, one fairly serious potential drawback to DirectQuery:

- Fetching data for multiple dashboard visuals can be unacceptably slow in some cases. This will depend on the data source, the network connection, and other elements.

- DirectQuery may work well when only one or two users are accessing a dashboard. However, it can grind to a halt when the number of users scales to the dozens–or hundreds.

Microsoft SQL Server Data

As a first example of DirectQuery at work, I will use a Microsoft SQL Server database as the data source. The steps are as follows:

1. Open a new Power BI Desktop application.

2. In the Power BI Desktop ribbon, click the small triangle at the bottom of the Get Data button and then click SQL Server. The SQL Server database dialog will appear.

3. Enter the server name in the Server text box. This will be the name of your SQL Server or one of the SQL Server resources used by your organization.

4. Enter the database name; if you are using the sample data, it will be CarSalesData.

5. Select the DirectQuery button. The dialog will look like Figure 4-1.

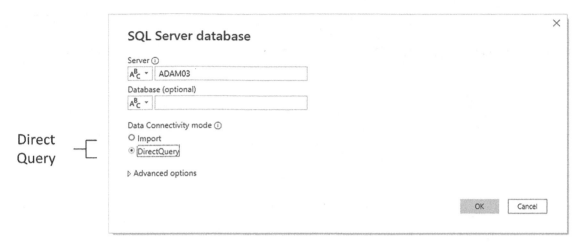

Figure 4-1. *The Microsoft SQL Server database dialog*

6. Click OK. The credentials dialog will appear. Define the type of credentials that you want to use, as you did in Chapter 3.

7. Click Connect. Power BI Desktop will connect to the server and display the Navigator dialog containing all the tables and views in the database that you have permission to see on the server you selected. In some cases, you could see a dialog saying that the data source does not support encryption. If you feel happy with an unencrypted connection, then click the OK button for this dialog.

8. Click the check boxes for the Clients, Colors, Countries, Invoices, InvoiceLines, and Stock tables. The data for the most recently selected data appears on the right of the Navigator dialog.

9. Click Load. Power BI Desktop will display the list of source tables for which it is establishing a *connection*. You can see this in Figure 4-2.

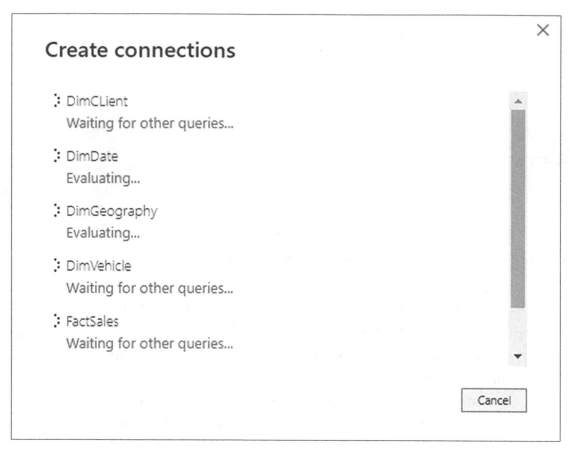

Figure 4-2. *Creating DirectQuery connections*

10. The Power BI Desktop window will open and display the tables
that you selected in the Fields list in the Report window. You can
see this in Figure 4-3.

Report and
Model Icons
Only

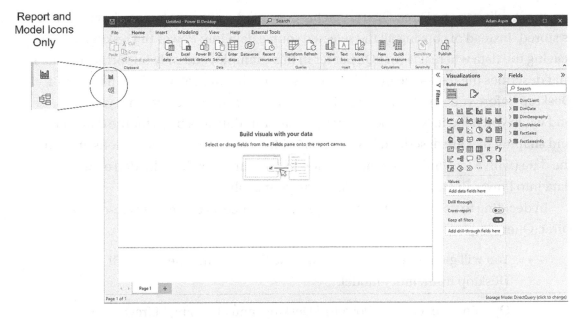

Figure 4-3. *Power BI Desktop using a direct connection*

This is so similar to the process that you saw as the first section of the previous chapter that you can be forgiven for asking, "So what's the difference?" Well there are a few differences, but they are so subtle as to be nearly invisible:

- There was no data load phase. When you clicked Load in step 9, the Power BI Desktop window appeared almost instantaneously. This is why the dialog that appeared briefly in step 9 says "Create Connections" instead of "Load Data."

- When you create Power BI Desktop visuals, they take longer to populate with data and display. For large source datasets, this delay can be considerable.

- To be slightly technical, what Power BI Desktop has done here is to query only the *metadata* from the source system. As metadata is nothing more than the description of the data and the data structures (or "data about data"), the process is extremely rapid as very little information is sent back from the server to Power BI Desktop. This is why establishing a DirectQuery connection is so fast, initially.

- You cannot access the underlying tables of data. This is why only the Report and Model icons are available.

By deciding to use DirectQuery, you have adopted a different logic to how the data is stored. Instead of copying all the selected source data into Power BI Desktop, you are leaving the data where it is (in SQL Server in this example) and only importing the data that describes the data—the metadata. So instead of needing all the data in Power BI Desktop before you can do any analysis, you can access only the data that you need, as and when you need it. Nevertheless, you can use this data as a basis for the data mashup and shaping that are described in Chapters 7–14. So you can add calculations and tweak the data (with a few limitations, at least) just as you can if you have loaded all the source data into Power BI Desktop before you begin your analysis.

In deciding to use DirectQuery, you have, in essence, accepted a trade-off. Using DirectQuery implies that

- You will gain time through *not* loading the data into the Power BI Desktop in-memory model.

- Opening the report in Power BI Desktop, and choosing Refresh, will update the fields in the model to reflect metadata changes.

- You will query the data source every time that you create, modify, or filter a Power BI Desktop visual—but *only* for the subset of the data that is required for the specific visual that you are creating or modifying.

- When you refresh the data, you will *not* reload all the data into memory as is the case when importing data. Power BI Desktop will only query the database for the *data that is actually required* to display the visuals that you have created.

- Slicing data can take longer when using DirectQuery because all the Power BI Desktop visuals are requeried.

- Some of the data mashup techniques that you will discover in Chapters 7–12 cannot be used. Fortunately, these are fairly rare occurrences.

- When the data necessary to service the request has recently been requested, Power BI Desktop uses recent data to reduce the time required to display the visualization. However, selecting Refresh from the Home ribbon will ensure all visualizations are refreshed with current data.

However, if you have a largely clean and coherent set of source data—such as data from a corporate data warehouse, where the "heavy lifting" required to make the data reliable and usable has already been carried out—then DirectQuery can enable access to very large data sources. And you are absolutely certain to be seeing the current data as well.

Note   You can equally well use a T-SQL query or a SQL Server stored procedure to return data over a DirectQuery connection. Simply expand the Advanced Options section of the connection dialog in step 5 (see Figure 4-1) and enter or copy the SQL text to execute as described in the previous chapter for loading data from SQL Server.

To give a balanced picture (and despite a heartfelt appreciation of the usefulness of DirectQuery), I have to admit that there are a few drawbacks to this connection type that you have to be aware of:

- The time taken to return queries from large databases can be extremely off-putting for users. They may have to wait an unacceptably long time to see the visuals loaded or updated.

- You cannot (for the moment at least) specify a stored procedure in the Advanced Options as the data source.

- The time required to refresh a visual is dependent on how long the data source takes to respond and return the results from the query.

- You can only return a maximum of one million records to Power BI Desktop when using DirectQuery. Fortunately, this threshold is a row limit, and not a limit on the source data. So if you are aggregating a billion-record data source but only returning 999,999 summary records, then the query will work.

- Really complex DAX queries simply will not work when using DirectQuery. The only solution to force complex queries to work is to switch back to loading the source data into the in-memory data model.

- Selecting File ➤ Options and Settings ➤ Options ➤ DirectQuery
  and then "Allow unrestricted measures in DirectQuery mode" will
  prevent Power BI Desktop from applying built-in limitations to DAX
  expressions. However, this can make some queries extremely slow, as
  the conversion from DAX to SQL is not always efficient.

- DAX Time Intelligence (as described in the companion volume "Pro
  DAX and Data Modeling") is not available with DirectQuery. Instead,
  more complex DAX functions have to be developed.

- Some capabilities in the Power BI Service (such as Quick Insights) are
  not available for datasets using DirectQuery.

- There is limited complexity available for SQL queries.

- Date/time support is only to one-second accuracy.

- Calculated columns are limited to being intrarow, as in, they can only
  refer to values of other columns of the same table, without the use of
  any aggregate functions.

- It is not possible to use the clustering capability to automatically
  find groups.

To end on a positive note, DirectQuery does come with the following advantages:

- Reports created using DirectQuery can, of course, be published to the
  Power BI Service.

- The 1GB limit on the dataset size in Power BI Desktop does not apply
  to DirectQuery connections.

Note   If Power BI Desktop has recently requested the data from the server that is
required for a visualization, then it will use the existing data that has been cached
to avoid placing undue stress on the source server as well as to enhance the user
experience. Consequently, you need to refresh the data if you want to be sure that
you are looking at the most up-to-date information and you suspect that the data
in the source has been updated recently.

SQL Server Analysis Services Dimensional Data

Another data source that can use DirectQuery (which uses the variant that Microsoft calls "Connect Live") is the SQL Server Analysis Services dimensional data warehouse— or "classic SSAS" as it is also known. Although a Live Connection to a classic SSAS dimensional data warehouse is very similar to loading data from SSAS into Power BI Desktop, there are a few differences that might make you prefer this method.

Note   Live Connection to a tabular data warehouse will only work if you are using SQL Server 2012 SP1 CU4 or greater. In this case, you have to have an Enterprise or Business Intelligence Edition unless you are using SQL Server 2016 or greater, in which case standard edition may be used.

Setting up a Live Connection to classic SSAS requires the following steps:

1. In the Power BI Desktop ribbon, click Get Data ➤ More ➤ Database and select SQL Server Analysis Services database in the Get Data dialog.

2. Click Connect. The SQL Server Analysis Services database dialog will appear.

3. Enter the Analysis Services server name and the database (or "cube") name, if you know it. In this example, the database is named CarSalesOLAP; of course, you will have to specify your own SSAS database name. In all cases, you will need to use the name of your own SSAS server.

4. Select the Connect live (or, in some cases, DirectQuery) button.

5. Click OK. If this is the first time that you are connecting to the cube, then the Access SQL Server Analysis Services dialog will appear so that you can define the credentials that you are using to connect to the Analysis Services database.

6. Accept or alter the credentials and click Connect. The Navigator dialog will appear.

7. Click the cube that you want to connect to from the SSAS
    database, drill into the cube elements, and select the metrics and
    dimensions that interest you. The dialog will look something like
    Figure 4-4.

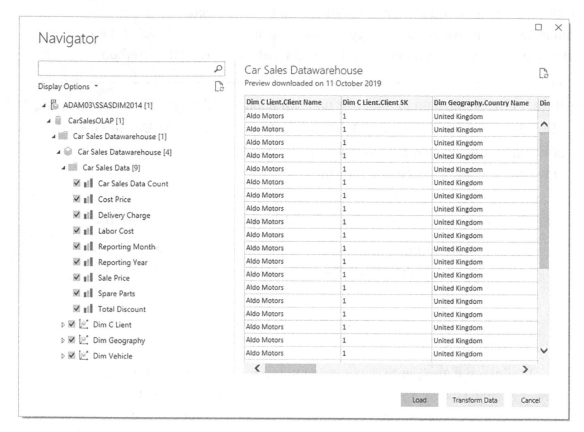

Figure 4-4. *Live Connect to a tabular database*

8. Click OK. The Power BI Desktop window will open and display the
    fact tables and dimensions that you selected in the Fields list in
    the Report window.

So although generally similar to the process for loading data from classic SSAS into
Power BI Desktop that you saw in the previous chapter, this approach does, nonetheless,
manifest one fundamental difference:

- You were not able to select the tables to use from the tabular data source. However, the underlying structure of the Analysis Services cube is visible just as it would be from, say, Excel. This includes visibility of the folder hierarchy for data that is present in the SSAS cube.

So, overall, a Live Connection implies that the source data *must* be ready to use and correctly structured for you to base your Power BI Desktop analytical reports on it. You *cannot* make changes to the data or the data structures or add any calculations in Power BI Desktop.

There are a couple of other points that you might need to take into account if you are hesitating between a Live Connection and loading data into Power BI Desktop:

- None of the data mashup possibilities that you will learn in Chapters 7–12 are available.

- You cannot use MDX to select the data that you want to use in Power BI Desktop. So a Live Connection is an "all-or-nothing" option.

- Certain features in your cube are either not supported fully or not supported at all with Live Connections from Power BI Desktop.

Yet, once again, a direct connection brings one crucial factor into the mix, and that is sheer speed when loading data. As SSAS data warehouses can be huge, the fact that you are not loading massive amounts of data into Power BI Desktop can save an immense amount of time when defining the source data. Not only that, a data warehouse that was too large to load into Power BI Desktop could now become accessible over a direct connection. Moreover, your Power BI Desktop visuals will only return the exact data that they need from the SSAS data source, as was the case for the SQL Server database in the previous section. All the hard work is carried out by the server, leaving Power BI Desktop (and you) free to concentrate on analysis and presentation.

However, the time taken to return data when refreshing dashboards may, in practice, become unworkable from a user experience perspective.

Microsoft SQL Server Analysis Services Tabular Data Sources

Now let's see how to use a Microsoft SQL Server Analysis Services tabular data warehouse as the data source for a Live Connection. A SQL Server Analysis Services tabular database is another technology that is used for data warehousing. It is different from the more traditional dimensional data warehouse in that it is entirely stored in the server memory and, consequently, is usually much faster to use.

To establish a Live Connection to an SSAS tabular data source:

1. In the Power BI Desktop ribbon, click Get Data ➤ More ➤ Database and select SQL Server Analysis Services database in the Get Data dialog.

2. Click Connect. The SQL Server Analysis Services database dialog will appear.

3. Enter the Analysis Services server name and the tabular database name (I don't tend to call these cubes). The database I am using is CarSalesTabular; you must specify your own tabular database name and the name of your own SSAS server.

4. Select the Connect live button. The dialog will look like Figure 4-5.

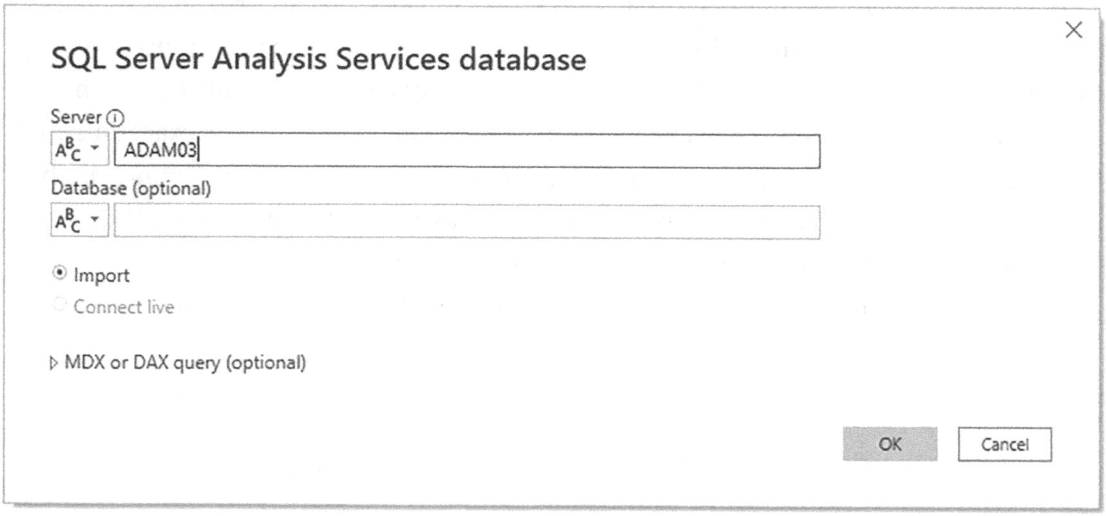

Figure 4-5. *Connecting to an SSAS (tabular) database*

5. Click OK. If this is the first time that you are connecting to the tabular data source, then the Access SQL Server Analysis Services dialog will appear so that you can define the credentials that you are using to connect to the Analysis Services database, where you will have to accept or alter the credentials and click Connect. The Navigator dialog will appear.

6. Click the perspective (Model in this example) that you wish to connect to. The dialog will look something like Figure 4-6.

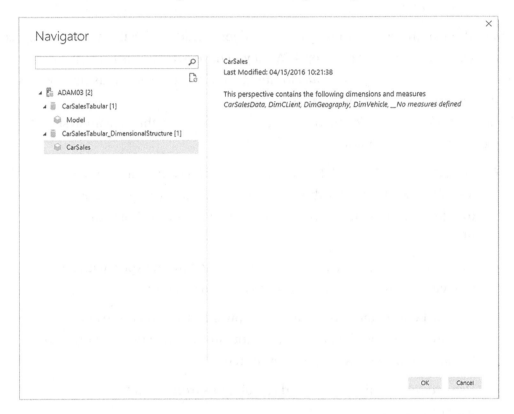

Figure 4-6. *Selecting attributes and measures from an SSAS tabular source*

7. Click OK. The Power BI Desktop window will open and display the tables that you selected in the Fields list in the Report window.

This process was fairly similar to the Connect Live connection that you established in the previous section. There are, nonetheless, a couple of further differences:

- The Relationships icon is no longer available on the top left of the Power BI Desktop window.

- None of the data mashup possibilities that you will learn in Chapters 7–14 are available.

- You were not able to select the tables to use from the tabular data source.

These three points actually imply the even deeper trade-off that you have accepted when you clicked Connect live in step 4. What you have agreed to is using the data source as the *complete and final model* for the data. You are, in effect, using Power BI Desktop as the front end for the data "as is."

This trade-off, however, is bursting with positives for you, the data analyst and dashboard creator. You can now

- See all the data in the data warehouse (or at least the data that you are authorized to see) without a complex process where you have to select dozens of tables (and risk overlooking a few vital sources of data).

- Access the data model as it has been designed by an expert with all the relationships between the tables defined at source.

- Get the latest data—what you see in Power BI Desktop is the current data in the tabular data warehouse. You only need to refresh Power BI Desktop if the data is refreshed at source.

- Use predefined hierarchies and calculations from the tabular data warehouse.

- *Access the data at lightning speed, because the source data is stored in the host server's memory*, and not on disk (as is the case with classic relational databases).

This last point is the one that really needs emphasizing. Not only are you gaining time through not loading a copy of the data into the Power BI Desktop in-memory data model, you are *accessing in-memory data on the server itself*, avoiding the need for slow searches for data on disk. The overall result is fast access to huge amounts of clean, structured, and aggregated data. All in all, the combination of a SQL Server tabular data warehouse and Power BI Desktop is designed to make data analysis considerably faster and easier.

To give a balanced overview, however, you need to understand that nothing in the Power BI ecosystem is quite as fast as data that is loaded into the Power BI Service. So while Connect Live to a tabular Analysis Service database is very fast and can enable access to very large datasets, the added latency of passing through the Power BI Service to a cloud or on-premises Analysis Services server will always slow things down to a greater or lesser extent.

It is worth noting that there is no real Power BI Desktop Query Editor access for a Live Connection to an Analysis Services tabular data warehouse. If you click the Transform data button, all you will see is the connection dialog for the Analysis Services database.

Note   When you set up a Live Connection to a tabular data source, you *cannot* filter the data that is loaded from an SSAS cube by expanding the MDX or DAX query (optional) item in the SQL Server Analysis Services database dialog and entering a DAX query.

DirectQuery with Non-Microsoft Databases

So far in this chapter, we have focused on using Microsoft data sources to establish DirectQuery connections to the source data. Fortunately, Microsoft has extended this technology to several other sources of corporate data.

Equally fortunate is the fact that connecting to (say) an Oracle database or a SAP HANA data warehouse using DirectQuery is every bit as simple as loading the data into Power BI Desktop from these (or indeed other) database and data warehouse sources. So I will not waste pages here in reexplaining the technique. If the source is a database, then you select DirectQuery. A DirectQuery connection will allow you to specify a SQL statement to select data, and a Live Connection will display all the objects in the data warehouse. It really is that simple.

DirectQuery and In-Memory Tables

A few years ago, the Microsoft SQL Server database was extended to include in-memory tables. These are perceived by tools like Power BI Desktop as standard tables, yet they exist in the server's memory (as opposed to storing the data on disk). This makes accessing data from in-memory tables extremely fast. When you add to this the fact that these tables can, in many cases, also use the compression technology that SSAS tabular data warehouses use, then you have pretty nearly the best of both worlds as far as analytics is concerned:

- Data stored in classic relational structures that is updated instantly

- Optimized structures for analytics (data is stored by column, rather than by row, and is both compressed and in-memory)

Here is not the place to expose all that this technology can bring to the table. However, as it is a potential game changer, it is worth showing you how you can use the latest versions of SQL Server (i.e., 2016 and up) as a direct source of analytical data. As this is essentially a rerun of the initial section of this chapter, I will not show here, again, the same screenshots of the process.

1. Open a new Power BI Desktop application.

2. In the Power BI Desktop ribbon, click the small triangle at the bottom of the Get Data button and then click SQL Server. The SQL Server database dialog will appear.

3. Enter the server name in the Server text box. This will be the name of your SQL Server or one of the SQL Server resources used by your organization.

4. Enter the database name; on my PC, it is CarSalesMemoryBased.

5. Select the DirectQuery button.

6. Click OK. The Access a SQL Server database dialog will appear. Define the type of credentials that you want to use, as you did in the previous chapter.

7. Click Connect. Power BI Desktop will connect to the server and display the Navigator dialog containing all the tables and views in the database that you have permission to see on the server you selected.

8.   Click the check boxes for the Clients, Colors, Countries, Invoices, InvoiceLines, and Stock tables. The data for the most recently selected data appears on the right of the Navigator dialog.

9.   Click Load. The Power BI Desktop window will open and display the tables that you selected in the Fields list in the Report window.

Yes, the process is identical to a "normal" DirectQuery connection. You will only return data when creating or modifying visuals. Yet here too, the data is stored in memory on the server, and all the heavy lifting is carried out by the server. However (and unlike when using a Live Connection to a tabular data warehouse), you can extend the data model, hide or add further columns, and create complex calculations. In many cases, this can be, quite simply, the best of all possible worlds.

DirectQuery and Refreshing the Data

You saw how to refresh data from databases and data warehouses in the previous chapter. If you are using DirectQuery, refreshing the data will likely only be necessary if the source data schema has changed. This is because every time that you create or modify a visual, or when you filter reports, pages, or visuals (or if you apply slicers), the data for any visual affected by the change is refreshed.

Consequently, you only have to click the Refresh button if you know that the source data schema has been updated. It could be, for instance, that the underlying database or data warehouse is reprocessed overnight—or even on an hourly basis. In cases like these, you should click the Refresh button once you know that the data schema has been updated on the server so that you can be absolutely sure that you are using the latest available data in your reports.

DirectQuery Optimization

In the case of a DirectQuery connection to a database, you could wish only to refresh selected tables. In this case, all you have to do is to right-click the table name in the Fields list and select Refresh for each table that you want to update.

Microsoft provides the following tips when using DirectQuery that you need to be aware of. Fortunately, they will probably not affect most standard use cases of DirectQuery:

- Ensure that the source database has been fully optimized and all necessary partitioning and optimal indexing has been applied.

- Avoid complex queries in Query Editor.

- Limit measures to simple aggregates.

- Avoid relationships on calculated columns.

- Avoid relationships on unique identifier columns (there is no unique identifier data type in Power BI Desktop), so this avoids a data type cast.

- Avoid calculated columns and data type changes.

- Assume referential integrity—to get inner joins.

- Do not use the relative data filtering in Query Editor.

Modifying Connections

If you are working in a structured development environment—or even if you are testing dashboards on a dataset that is either an old version or possibly on a nonlive server—you could want at some point to switch from a current data source to another source. Power BI Desktop will let you do this. However, switching data sources will only work if the *structures of the source and the destination data are identical*. Practically, this means that the server and database can be named differently, but the tables and fields must have the same names.

To see this in action, you can do the following:

1. In an existing Power BI Desktop file, click the small triangle at the bottom right of the Transform data button in the Home ribbon, and select Data source settings. The Data source settings dialog will be displayed. Figure 4-7 shows you this dialog for the SQL Server connection that you saw at the start of this chapter.

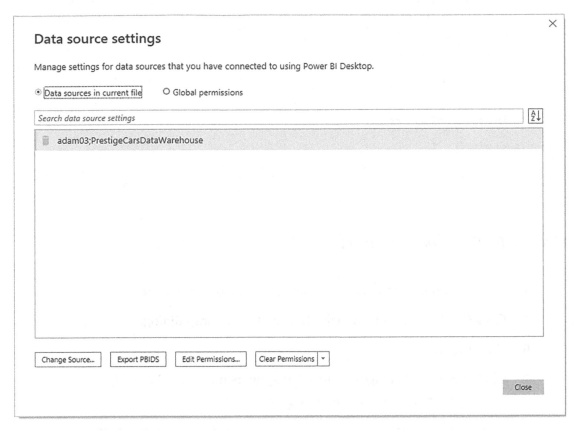

Figure 4-7. *The Data source settings dialog*

2. Click the connection that you want to modify.

3. Click Change Source. The dialog that originally allowed you to
    specify the data source (in this example, that is the SQL Server
    database connection dialog) will appear. You can see this in
    Figure 4-8.

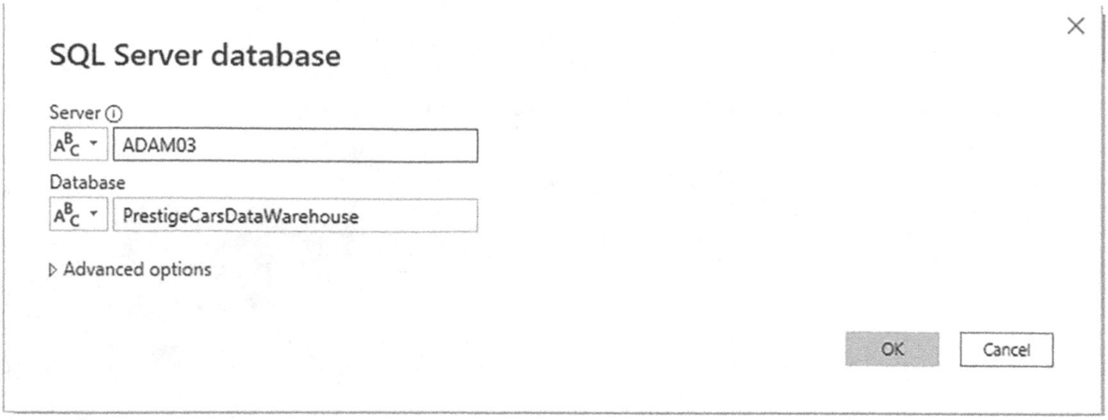

SQL Server database

Server ⓘ

A^B_C ▾ | ADAM03

Database

A^B_C ▾ | PrestigeCarsDataWarehouse

▷ Advanced options

OK   Cancel

Figure 4-8. *The connection dialog*

4. Specify a different server and/or database as the data source.

5. Click OK. You will return to the Data source settings dialog.

6. Click Close.

7. Click the Apply Changes button that appears under the Power BI Desktop ribbon at the top of the screen.

Note   If this is the first time that you are establishing a connection to this new server and database, you will have to specify the credentials to use.

Assuming that the new data source contains the same table names and structures, the existing data will be replaced with the data from the new source that you specified.

I have to stress again that this technique will only work if the underlying database metadata is *identical* across the two servers. If the data structures are not the same, you will see an error dialog similar to that shown in Figure 4-9.

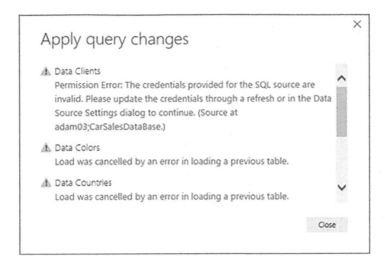

Figure 4-9. *Modifying a database connection*

In cases like this, you may well have to rebuild a new Power BI Desktop file using the new data source. This could mean re-creating or copying any data mashups and formulas (as well as actual visualizations) from the old version to the new file.

Changing Permissions

It is all too frequent when working with databases and data warehouses to encounter permissions problems. It could be that you set up a connection to a database which required you to change the password at a later date. Meanwhile, the password stored in Power BI Desktop is the old version. So when you try to update your dashboard, you hit a blocker.

Fortunately (assuming, at least, that you know the new password), you can update your stored credentials in Power BI Desktop. As an example, let's suppose that you want to update your Oracle password. Here is how:

1. In an existing Power BI Desktop file, click the small triangle at the bottom right of the Transform data button in the Home ribbon, and select Data source settings.

2. Click the Global permissions radio button at the top of the dialog. The Data source settings dialog will be displayed. Figure 4-10 shows you this dialog for the current Power BI Desktop connections on my laptop.

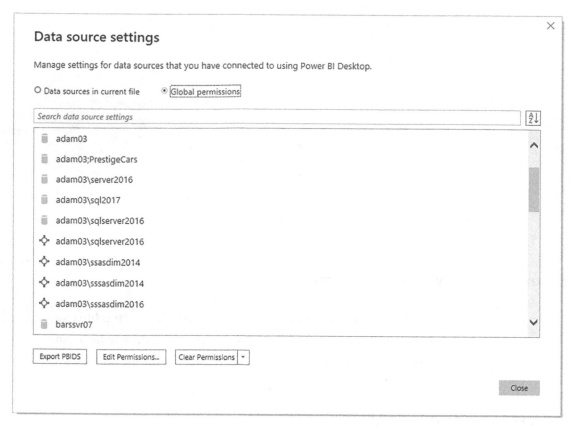

Figure 4-10. *Permissions for current connections*

3. Click the name of the connection that you want to change (ADAM03 in this example).

4. Click the Edit Permissions button. The Edit Permissions dialog will be displayed, as shown in Figure 4-11.

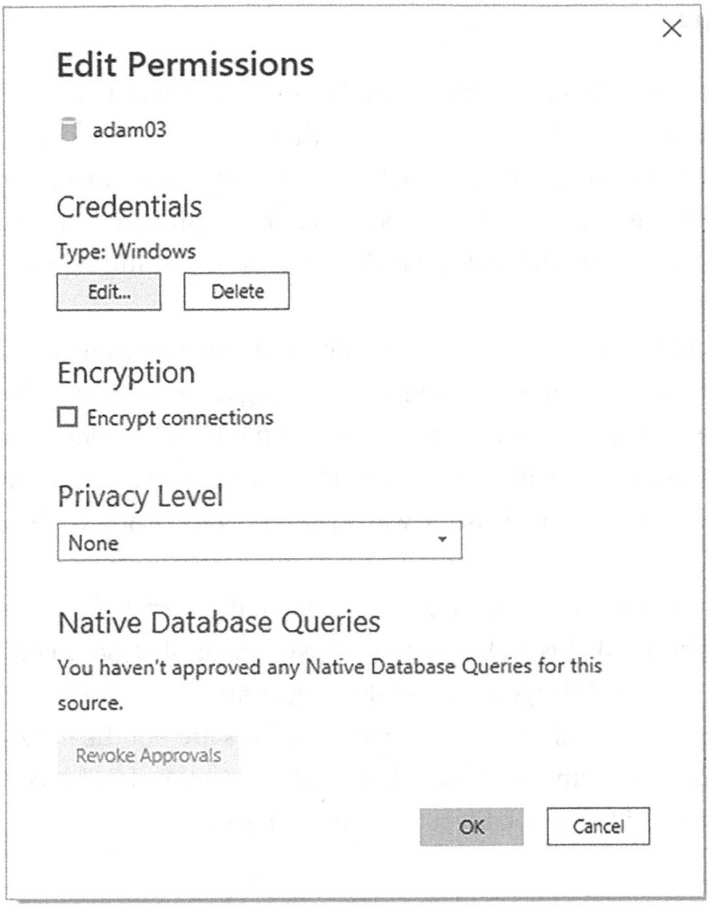

Figure 4-11. *The Edit Permissions dialog*

5. Click Edit. The connection permissions dialog will be displayed (shown earlier in Figure 3-14 in the section that explains how to connect to an Oracle database).

6. Modify the password (or even the username if this is required).

7. Click Save. This returns you to the Edit Permissions dialog.

8. Click OK. You will return to the Data source settings dialog.

9. Click Close to return to Power BI Desktop.

You can now refresh the data using the new permissions that you have just entered.

Conclusion

This short chapter extended your abilities to use databases and data warehouses as the data source for your Power BI analysis. You saw that you can choose not to load data into the Power BI Desktop in-memory data model but can, instead, connect directly to the source data repository. Moreover, you saw that the approach to setting up a direct connection is virtually identical to the way that you set up a standard data load into Power BI Desktop.

DirectQuery allows you to avoid a potentially massive data transfer. This can save you a large amount of time when loading or refreshing data and also guarantees that you are looking at the latest data. So while it can take a little longer to design and filter your visuals, you are only using the precise data that you need on each and every occasion. Sometimes, this can be the only feasible way to analyze data from terabyte-sized data warehouses.

When combined with in-memory data storage on the server (in in-memory tables or tabular data warehouses), this technique can make Power BI Desktop into a compelling front end for the analysis of huge and complex corporate data sources.

However, corporate databases and data warehouses are not the only sources of large-scale data that exist. More and more data is now stored in the cloud. Accessing cloud- and web-based data will be the subject of the next chapter.

CHAPTER 5

Loading Data from the Web and the Cloud

In this chapter, we will take a look at a subset of the fast-growing and wide-ranging set of data sources available over the Internet that you can use as a source of analytical data for Power BI Desktop. While the data sources that you will see in the following pages may be extremely diverse, they all have one thing in common: they are stored outside the enterprise and are available using an Internet connection.

The ever-increasing range of data sources that are available are provided by a multitude of suppliers. Looking at all the current sources would take up an entire book, so I will show you how to access some of the mainstream services that are currently on offer. Once you have learned how to access a few of them, you should be able to extend the basic techniques to access just about any of the web and cloud services that can currently be used by Power BI Desktop.

Power BI Desktop is now firmly entrenched as a fundamental part of the Microsoft universe. As PowerBI.com (the cloud service that you can use to store and share dashboards) is part of the Microsoft cloud offering, it is perhaps inevitable that the Power BI Desktop developers have gone out of their way to ensure that Power BI has become the analytical tool of choice for solutions that are hosted in Azure. For this reason, I will explain quite a few of the core services that host data in Azure–the Microsoft cloud.

Nearly all of the data connections outlined in this chapter require access to a specific online source. Most of these sources are industrial strength—and not free. However, if your enterprise is not a subscriber to these services, and you wish, nevertheless, to experiment with them, it could be worth taking a look at the free trial offers available from many (if not all) of the service providers whose offerings are outlined in this chapter.

© Adam Aspin 2022
A. Aspin, *Pro Data Mashup for Power BI*, https://doi.org/10.1007/978-1-4842-8578-7_5

Web and Cloud Services

Before delving into the details of some of the web and cloud services that are available, let's take an initial high-level look at what these really are. These data sources include (among many others)

- Web pages

- Online services, such as Google Analytics, Salesforce, and MS Dynamics 365

- Microsoft Azure, which covers hosting files in Azure Blob services, storing data in an Azure SQL database, or accessing Azure Synapse Analytics (or reading big data in other platforms)

- OData, a generic method of accessing data on the Internet

Web Pages

If you need to collect some data that you can see as a table in a web browser, you can use Power BI Desktop to connect to the URL for the page in question and then load all the data from any table on the page.

Online Services

Online services is a catch-all phrase used to describe data that you can access using the Internet. Most of the online services available to Power BI Desktop are what are called "platforms." These are (often huge) software and data resources that either are only available online or were once housed in corporate systems but are now available as services on the Internet. There are currently dozens of online services that are available to connect to using Power BI Desktop. Indeed, the number of available services is growing at a startling pace. Some of the more frequently used include those listed in Table 5-1.

Table 5-1. *Online Services Currently Available to Power BI Desktop*

Source	Comments
Azure DevOps	Accesses Azure DevOps data entities.
Salesforce Objects	Lets you access data in Salesforce.
Salesforce Reports	Lets you access the prestructured data objects (both native and custom) that underlie built-in Salesforce reports.
Google Analytics	Lets you access the data managed by Google to track website traffic.
Adobe Analytics	Lets you access the data tracked in Adobe Analytics.
SharePoint Online	Connects to the cloud-hosted version of Microsoft SharePoint.
Microsoft Exchange Online	Connects to the cloud-hosted version of Microsoft Exchange.
Dynamics 365 Online	Connects to the cloud-hosted version of Microsoft Dynamics 365—the MS CRM and ERP solution.
Dynamics 365 Business Central	Connects to Dynamics 365 Business Central (online and on-premises).
Dynamics NAV	Connects to Microsoft Dynamics NAV.
OData	Although OData is not, technically, an online platform, it is certainly an online source of data. This is a standardized method for connecting to different data structures using a URL as a starting point.
GitHub	Connects to GitHub.

These are only some of the many available online services that you can connect to in Power BI Desktop. I advise you to examine the list and see if the service that you require is currently available. Remember also that this list is growing all the time. So don't hesitate to check it frequently if you have a cloud-based product that you need to connect to Power BI.

Note   As the number of available online services is increasing at an ever-increasing rate, you will probably find many more than those that I have listed here by the time that this book is published. Moreover, there are currently a number of online services that are available as beta versions. This means that you can test them, but they are not yet finalized and supported.

Microsoft Azure

Azure is the Microsoft cloud. The Azure data sources that Power BI Desktop can currently connect to, and can preview and load data from, are given in Table 5-2.

Table 5-2. *Azure Sources*

Source	Comments
Microsoft Azure SQL database	Lets you connect to a Microsoft SQL Server cloud-based database and import records from all the data tables and views that you are authorized to access
Azure Synapse Analytics (formerly Azure SQL Data Warehouse)	Lets you connect to Microsoft's cloud-based, elastic, enterprise data warehouse and Serverless SQL Pool
Azure Analysis Services database	Connects to an Azure Analysis Services database
Microsoft Azure Blob Storage	Reads from a cloud-based unstructured data store
Microsoft Azure Table Storage	Reads from Microsoft Azure tables
Microsoft Azure Cosmos DB	Lets you connect to Microsoft's Azure-hosted NoSQL database
Microsoft Azure Data Lake Storage (Gen1 and Gen2)	Lets you connect to Microsoft's big data analytics storage
Microsoft Azure HDInsight	Reads cloud-based Hadoop files in the Microsoft Azure environment
Azure HDInsight Spark	Lets you connect to Microsoft's HDInsight parallel processing framework in Microsoft cloud
HDInsight Interactive Query	Runs interactive queries over HDInsight
Azure database for PostgreSQL	Accesses Azure database for PostgreSQL
Azure Databricks	Lets you connect to Databricks in Azure
Azure Synapse Analytics workspace	Lets you connect to Azure Synapse Analytics workspaces in Azure

Obviously, more Azure connection options are currently available–and more are being added to Power BI Desktop by Microsoft as the Azure offering is extended.

Web Pages

As a first and extremely simple example, let's grab some data from a web page. Since I want to concentrate on the method rather than the data, I will use a web page that has nothing to do with the sample data in the book. I will not be using this other than as a simple introduction to the process of loading data from web pages using Power BI Desktop.

Assuming that you have launched Power BI Desktop and closed the splash screen:

1. Click the small triangle at the bottom of the Get Data button in the Home ribbon.

2. Select Web from the menu that appears, as shown in Figure 5-1.

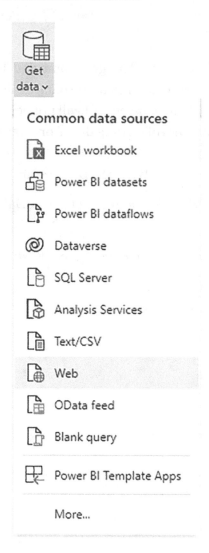

Figure 5-1. *The Get Data menu*

3. Enter the following URL (it is a Microsoft help page for Power
    BI Desktop that contains a few tables of data): http://office.
    microsoft.com/en-gb/excel-help/guide-to-the-power-
    query-ribbon-HA103993930.aspx. I am, of course, hoping that it
    is still available when you read this book. Of course, if you have a
    URL that you want to try out, then feel free! The dialog will look
    something like Figure 5-2.

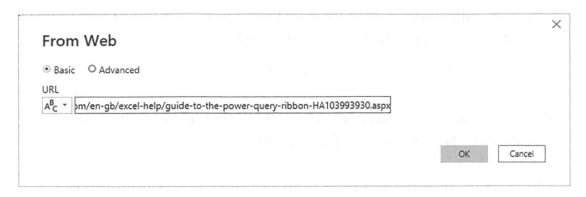

Figure 5-2. *The From Web dialog*

4. Click OK. The Navigator dialog will appear. After a few seconds, during which Power BI Desktop is connecting to the web page, the list of available tables of data in the web page will be displayed.

5. Click one of the table names on the left of the Navigator dialog. The contents of the table will appear on the right of the Navigator dialog to show you what the data in the chosen table looks like, as shown in Figure 5-3.

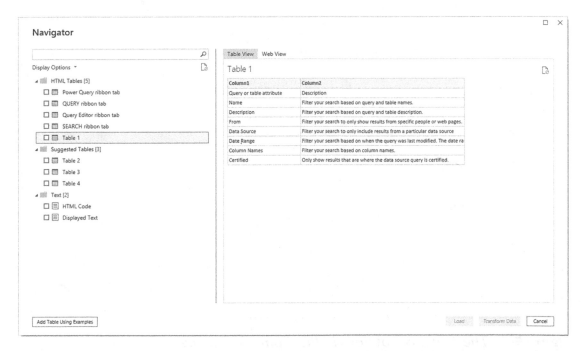

Figure 5-3. *The Navigator dialog previewing the contents of a table on a web page*

6. Select the check box in the Navigator dialog (shown to the left of Table 4 in Figure 5-3). You can select multiple tables if you wish.

7. Click Load at the bottom of the window.

Tip   Another way of accessing web pages is to click Get Data ➤ Other. You can then select Web in the list on the right of the Get Data dialog.

This simple example showed how you can load tables of data from a web page and load it into Power BI Desktop.

Advanced Web Options

In step 3 of the previous example, you could have selected the Advanced button. Had you done this, the From Web dialog would have expanded to allow you to build complex URLs by adding URL parts. You can see an example of this in Figure 5-4.

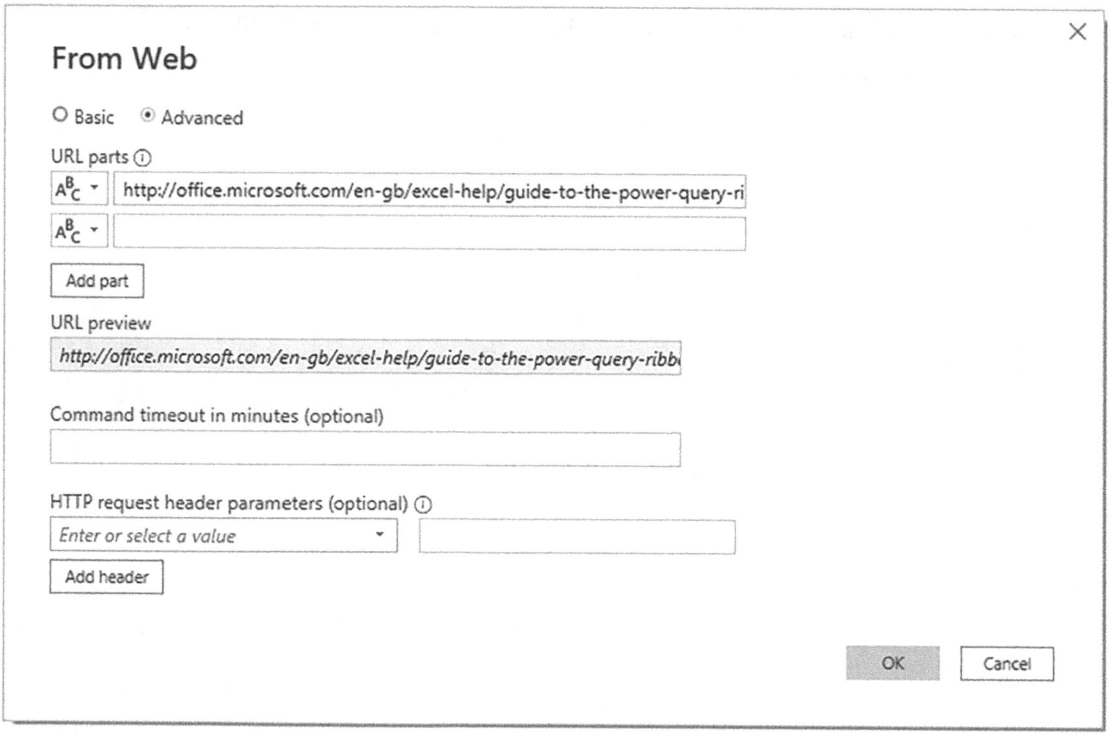

Figure 5-4.   *The Advanced options in the From Web dialog*

Clicking the Add part button allows you to define multiple URL parts.

If necessary, you can also specify HTTP request header parameters that will be used when submitting the URL. These could be required by certain web pages. A discussion of these is outside the scope of this book.

Table View or Web View

Looking at the tables that a web page contains is not always the most natural way of finding the right data. This is because you are looking at the data tables out of context. By this, I mean that you cannot see where they are on the web page. After all, the Web is a very visual medium.

To help you find the correct data table on a web page, the Query Editor lets you switch between two views of the web source:

- Table view (which you saw in Figure 5-3)

- Web view (which you can see in Figure 5-5)

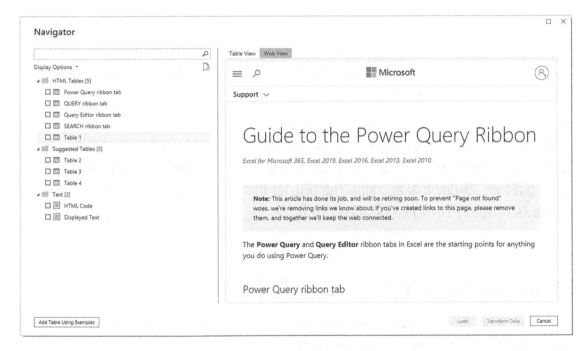

Figure 5-5. *Web View in the Navigator dialog*

In essence, the Navigator now includes a "mini web browser" so that you can see the tables in the context of the web page. You alternate between these ways of visualizing the web page by clicking the Table View and Web View buttons that are at the top center of the Navigator dialog. The same web page that you saw previously looks like Figure 5-5 when you switch to Web View.

Salesforce

One of the pioneers in the Software-as-a-Service (SaaS) space—and now, indisputably, one of the leaders—is Salesforce. So it is perhaps inevitable that Power BI Desktop will allow you to connect to Salesforce and load any data that you have permission to view using your Salesforce account.

Indeed, Salesforce is such a wide-ranging and complete service that you have two possible methods of accessing your data:

- Objects

- Reports

Briefly, Salesforce objects are the underlying data tables that contain the information that you want to access. Salesforce reports are the data that has been collated from the data tables into a more accessible form of output.

Tip   If you do not have a corporate Salesforce account but want, nevertheless, to see how to use Power BI Desktop to connect to Salesforce data, you can always set up a free 30-day trial account. The URL for this is `www.salesforce.com/form/signup/freetrial-sales.jsp`.

Loading Data from Salesforce Objects

Assuming, then, that you have a valid Salesforce account, here is how you can load data from Salesforce objects into Power BI Desktop:

1. In the Power BI Desktop Home ribbon, click the Get Data button.

2. Click Online Services on the left, and then select Salesforce Objects on the right. The Get Data dialog will look like Figure 5-6.

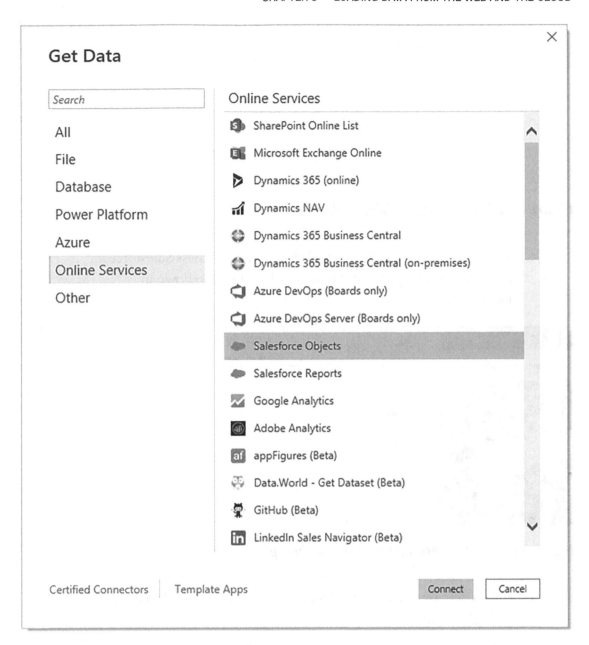

Figure 5-6. *The Get Data dialog for online services*

3. Click Connect. The Salesforce Objects dialog will appear. It should look like the one shown in Figure 5-7.

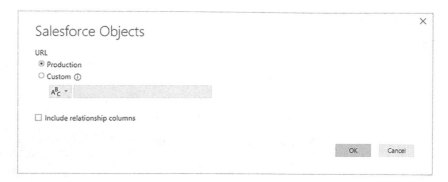

Figure 5-7. *The Salesforce Objects dialog*

4. Select the Production button and click OK. The Access Salesforce
    login dialog will appear. It should look like the one shown in
    Figure 5-8.

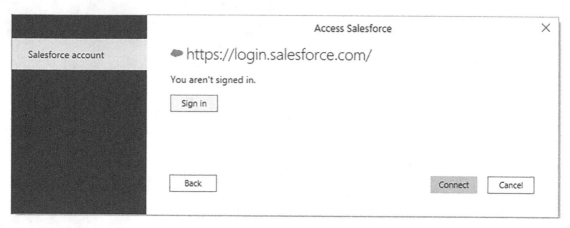

Figure 5-8. *The Access Salesforce login dialog*

5. Unless you are already signed in, click Sign in. The Salesforce sign-
    in dialog will appear.

6. Enter your Salesforce login and password. The dialog should look
    something like the one shown in Figure 5-9.

Figure 5-9. *The Salesforce sign-in dialog*

7. If this is the first time that you are connecting to Salesforce from
    Power BI Desktop (or if you have requested that Salesforce request
    confirmation each time that you log in), you will be asked to
    verify your identity. The Salesforce Verify Your Identity dialog will
    appear, as shown in Figure 5-10.

Figure 5-10.   *The Salesforce Verify Your Identity dialog*

8.   Click Verify. Salesforce will send a verification code to the email
     account that you are using to log in to Salesforce.

9.   Enter the code in the Verification Code field and click OK. You will
     see the Allow Access dialog, as in Figure 5-11.

Figure 5-11. *The Salesforce Allow Access dialog*

10. Click Allow. You will return to the Access Salesforce dialog, only now you are logged in. You can see this in Figure 5-12.

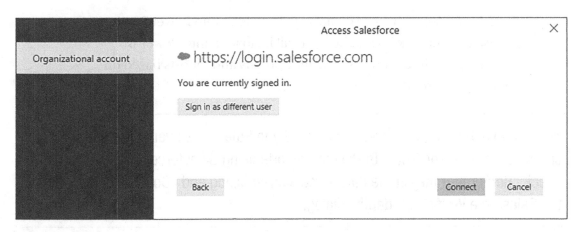

Figure 5-12. *The Access Salesforce dialog*

11. Click Connect. The Navigator will appear, showing the Salesforce objects that you have permissions to access. You can see an example of this in Figure 5-13.

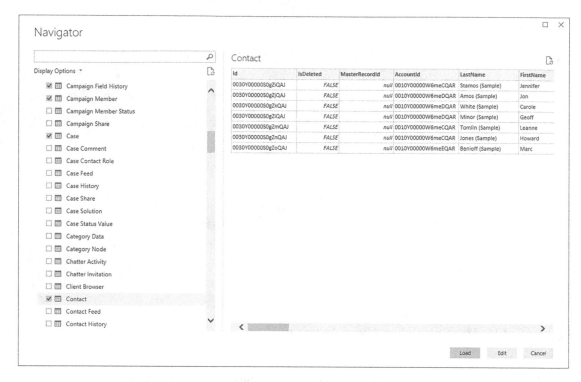

Figure 5-13. *Salesforce objects viewed in the Navigator*

12. Select the objects whose data you wish to load into Power BI Desktop and click Load. The data will be loaded into Power BI Desktop ready for you to create dashboards and reports based on your Salesforce data.

Tip   To avoid having to confirm your identity to Salesforce every time that you create a new suite of Power BI Desktop reports using Salesforce data, you can check "Remember me" in the Salesforce sign-in dialog and "Don't ask again" in the Salesforce Verify Your Identity dialog.

Salesforce objects contain a vast amount of data. However, from the point of view of Power BI Desktop, this is similar to accessing a database structure. This means that you have to have some understanding of how the underlying data is stored. Should you wish to learn about the way that Salesforce data is structured, then I suggest that you start with the Salesforce documentation currently available at `https://trailhead.salesforce.com/en/content/learn/modules/data_modeling/objects_intro`.

Note   It is not currently possible to write SOQL queries to filter the data loaded from Salesforce into Power BI desktop.

Salesforce Reports

If you find that you are simply submerged by the amount of data that is available in Salesforce, you can, instead, go directly to the data that underlies standard Salesforce reports. This will avoid your having to learn about the underlying data structures. The downside is that you cannot easily extend these datasets.

To access Salesforce report data, simply follow the steps outlined in the previous section. However, instead of choosing Salesforce Objects in step 2, select Salesforce Reports instead. The Navigator dialog will, in this case, look something like the one shown in Figure 5-14.

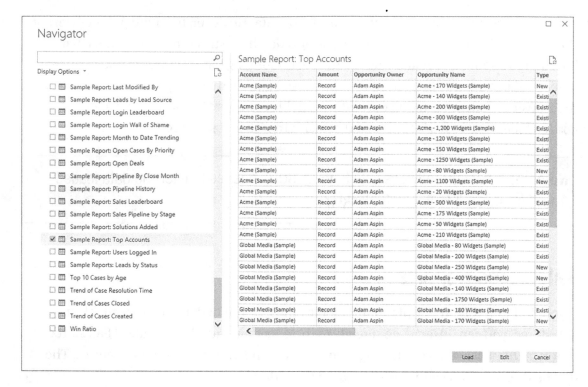

Figure 5-14. *The Navigator dialog showing the data for Salesforce Reports*

From here, you can select and load the reports data from Salesforce that you want to use to create your own dashboards.

Microsoft Dynamics 365

Another online service that contains much valuable enterprise data is Microsoft Dynamics 365. As you would probably expect, Power BI Desktop can connect easily to Microsoft online sources such as Dynamics. Here is how to do this:

Tip   If you do not have a corporate Microsoft Dynamics 365 online account but want, nevertheless, to see how to use Power BI Desktop to connect to Microsoft Dynamics 365 data, you can always set up a free 30-day trial account. The URL for this is currently `https://dynamics.microsoft.com/en-gb/dynamics-365-free-trial/`. Indeed, this example is from using a free 30-day trial account (which will likely have expired long before this book is in print).

---

1. In the Power BI Desktop Home ribbon, click the Get Data button.

2. Click Online Services on the left, and then select Dynamics 365 (online) on the right. The Get Data dialog will look like Figure 5-15.

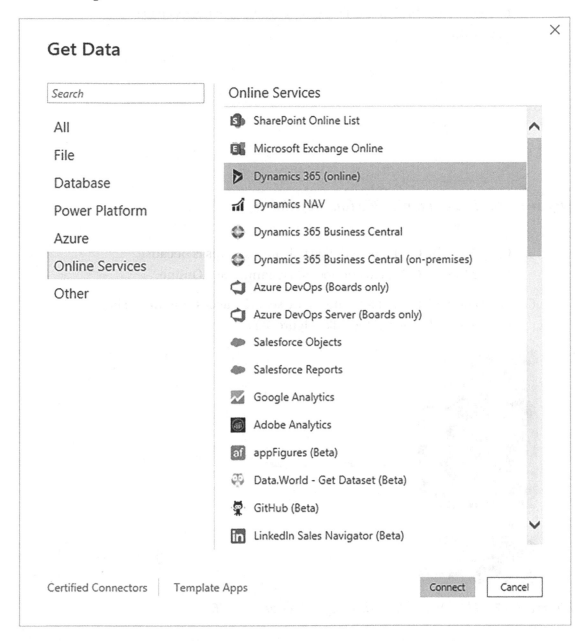

Figure 5-15. *The Get Data dialog for Dynamics 365*

3. Click Connect. The Dynamics 365 (online) dialog will appear.

4. Enter the URL that you use to connect to Dynamics 365 and add
    /api/data/v8.1 (at least, this was the case as this book went
    to press). It could look like the one shown in Figure 5-16. Note,
    however, that this URL will vary depending on where you are in
    the world.

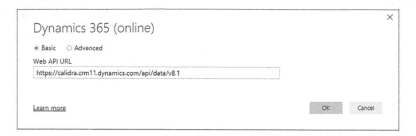

Figure 5-16.   *The Dynamics 365 (online) dialog*

5. Click OK. The OData feed dialog will appear. This is because
    Power BI uses OData to connect to Dynamics 365 Online.

6. Select Organizational account as the security access method. The
    OData feed dialog will look like Figure 5-17.

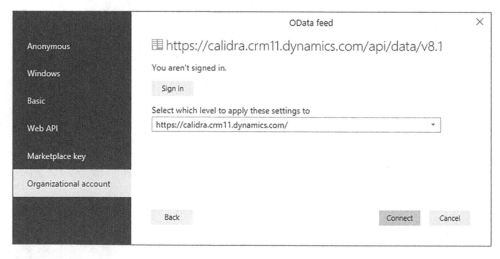

Figure 5-17.   *The OData feed dialog for Dynamics 365*

7. Click Sign in to sign in to your Dynamics 365 account and follow
    the Microsoft sign-in process. Once completed, the OData feed
    dialog will look something like the one in Figure 5-18.

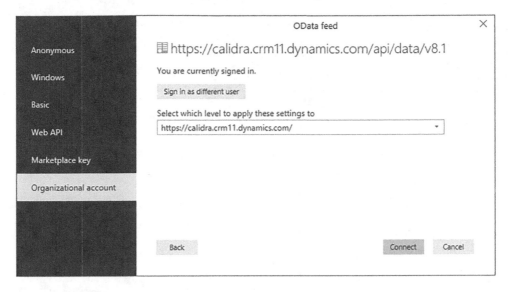

Figure 5-18. *The OData feed dialog for Dynamics 365 after sign-in*

8. Click Connect. The Navigator dialog will appear showing all the
    Dynamics objects that you have permissions to connect to. You
    can see an example of this in Figure 5-19.

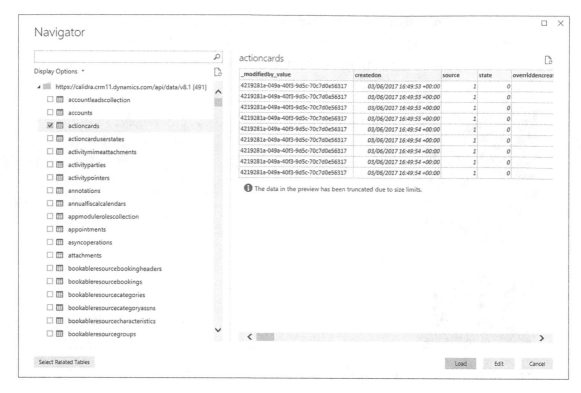

Figure 5-19. *The Navigator dialog for Dynamics 365*

Note   In step 6, you saw that an MS Dynamics 365 connection is really an OData connection. OData is explained in more detail in a subsequent section of this chapter.

There are a huge number of Dynamics 365 tables—and this number will vary depending on the subscription that your organization has taken out. However, you are, in reality, accessing a database structure. This means that you have to have some understanding of how the underlying data is stored. Should you wish to learn about Dynamics 365 tables, then I suggest that you start with the Microsoft online help at `https://docs.microsoft.com/en-us/dynamics365/fin-ops-core/dev-itpro/data-entities/data-entities`.

Google Analytics

Assuming that you have a valid Google Analytics account set up, you can use Power BI Desktop to connect to the Google Analytics data that you have permissions to access. For this example to work, you will need a valid and functioning Google Analytics account.

Note   To sign up for a Google Analytics account that you can use to test Power BI Desktop, go to `www.google.com/analytics`.

---

1. In the Power BI Desktop Home ribbon, click the Get Data button.

2. Click Online Services on the left, and then select Google Analytics on the right. The Connecting to a third-party service dialog will look like Figure 5-20.

Figure 5-20. *The third-party service connector alert*

3. Click Continue. The Google Account dialog will appear. This currently looks like the one in Figure 5-21.

Figure 5-21. *The Google Analytics connection dialog*

4. Click Sign in. The Google Choose an account dialog will be displayed. This currently looks like the image in Figure 5-22.

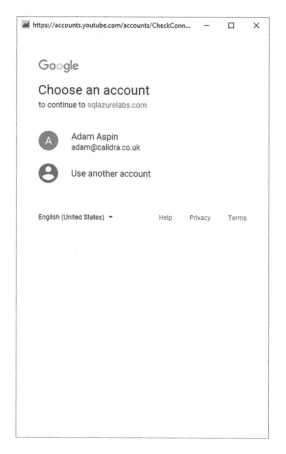

Figure 5-22. *The Google Analytics login dialog*

5. Click the existing account to use for Google Analytics. The Google
   Analytics permissions dialog will appear. This currently looks like
   the one in Figure 5-23.

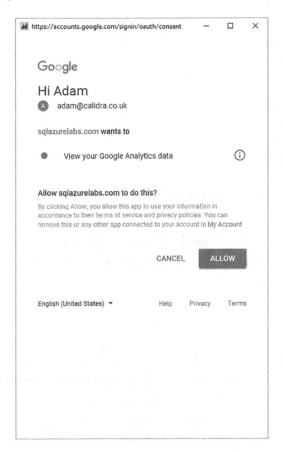

Figure 5-23. *The Google Analytics permissions dialog*

6. Click Allow. You will return to the Google Account dialog, but
    logged in this time. You can see this in Figure 5-24.

Figure 5-24. *The Google Analytics dialog when signed in*

7. Click Connect. The Navigator dialog will appear, displaying the
    data tables that you can connect to in Google Analytics.

OData Feeds

OData is a short way of referring to the Open Data Protocol. This protocol allows web
clients to publish and edit resources, identified as URLs. The data that you connect to
using OData can be in a tabular format or indeed in different structures.

OData is something of a generic method of connecting to web-based data.
Consequently, each OData source could differ from others that you may have used
previously. Indeed, you have already seen OData when connecting to Dynamics 365.

However, there are a multitude of OData sources that are available. Some are public;
some are only accessible if you have appropriate permissions. However, the access
method will always be broadly similar. Here, then, is an example of how to connect to an
OData sample source that Microsoft has made freely available (at least when this book
went to press):

1. In the Power BI Desktop Home ribbon, click the small triangle
    at the bottom of the Get Data button. Alternatively, click the
    Get Data button and select Other from the left pane of the Get
    Data dialog.

2. Select OData feed from the menu. The OData feed dialog will appear.

3. Enter the URL that you are using to connect to the OData source. In this example, I will use a Microsoft sample OData feed that you can find at `http://services.odata.org/northwind/northwind.svc`. The dialog should look like Figure 5-25.

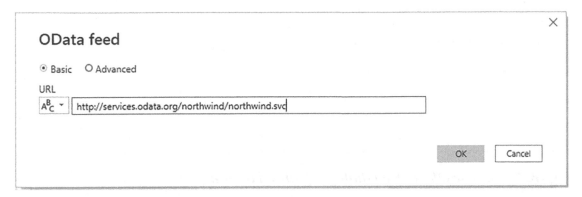

Figure 5-25. *The OData feed dialog*

4. Click OK. The Navigator dialog will be displayed and will show the data available using the specified URL.

OData Options

The OData feed dialog (rather like the From Web dialog that you saw earlier in this chapter) also contains an Advanced button. Selecting this will expand the dialog to allow you to add one or more URL parts to the URL. You can see this in Figure 5-26.

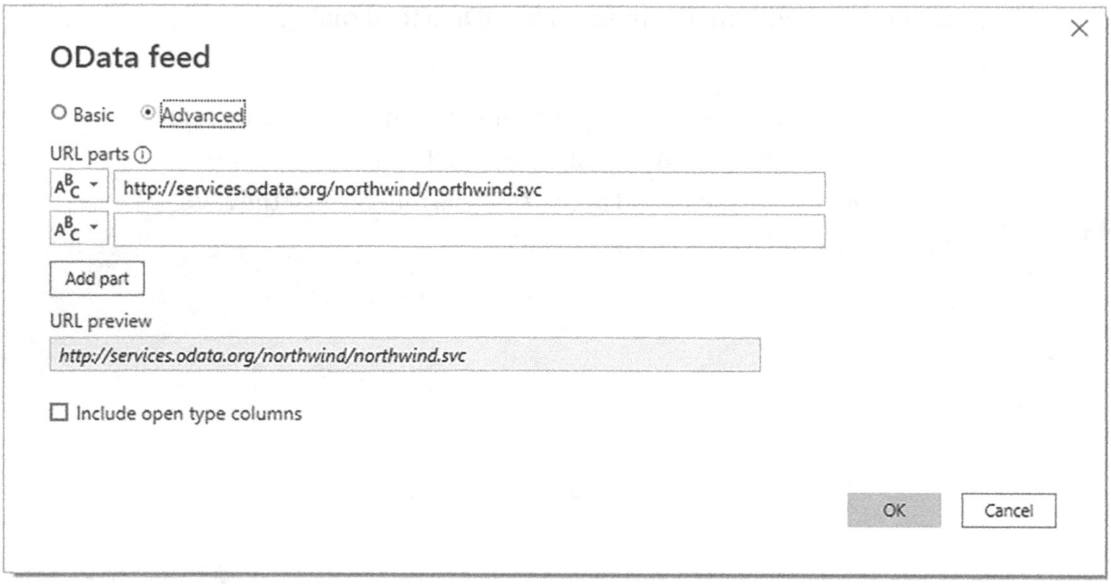

Figure 5-26. *The OData feed dialog Advanced options*

Note   URL parts can be parameterized in the Power BI Desktop Query Editor. I will explain parameterization in Chapter 13.

Azure SQL Database

SQL Server does not only exist as an on-premises database. It is also available as a "Platform as a Service" (also known as PaaS). Simply put, this lets you apply a pay-as-you-go model to your database requirements where you can fire up a database server in the cloud in a few minutes and then scale it to suit your requirements, rather than buying hardware and software and having to maintain them.

Connecting to Microsoft's PaaS offering, called Azure SQL database, is truly simple. If you have the details of a corporate Azure SQL database, you can use this to connect to it. If you do not, and nonetheless want to experiment with connecting Power BI Desktop to Azure SQL database, you can always request a free trial account from Microsoft and set up an Azure SQL database in a few minutes. If this is the path that you are taking, then you can find instructions on how to do this (including loading the sample data that you will connect to later in this section) at the following URL: `https://docs.microsoft.com/en-gb/azure/azure-sql/database/single-database-create-quickstart`.

Tip   When you are creating an Azure SQL database for test purposes, be sure to define the source to be *Sample*. This will ensure that the MS sample data is loaded into your test database.

To connect from Power BI Desktop to an Azure SQL database:

1. Open a new Power BI Desktop application.

2. In the Power BI Desktop ribbon, click the small triangle at the bottom of the Get Data button and then click More.

3. Click Azure in the list on the left and then Azure SQL database on the right. The Get Data dialog will look like the one in Figure 5-27.

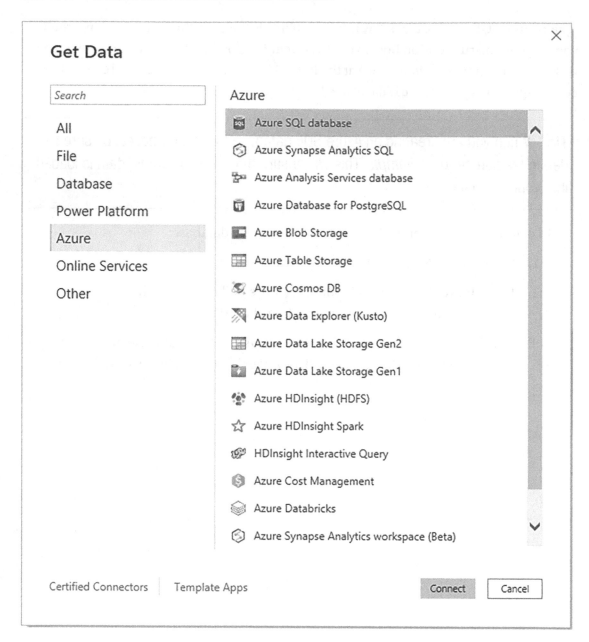

Figure 5-27. *Azure data sources in the Get Data dialog*

4. Click Connect. The SQL Server database dialog will appear (after all, an Azure SQL database is a SQL Server database—but in the cloud).

5. Enter the Azure SQL database server name that you obtained from the Microsoft Azure management portal (or that was given to you by a corporate DBA). The SQL Server database dialog will look like the one shown in Figure 5-28.

Figure 5-28. *The SQL Server database dialog for an Azure SQL database connection*

6. Click OK. The credentials dialog will appear.

7. Click Database on the left and enter a valid username and password. The credentials dialog will look like the one shown in Figure 5-29.

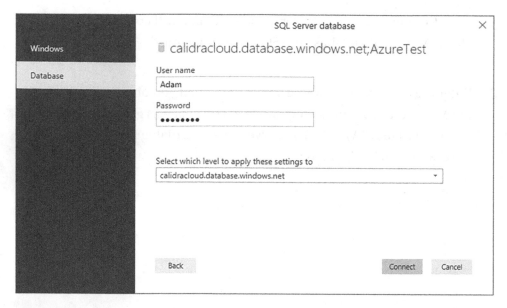

Figure 5-29. *The SQL Server credentials dialog for an Azure SQL database connection*

8. Click Connect. The Navigator dialog will appear showing the database(s) that you have permission to access in the Azure SQL Server database. This dialog will look like the one shown in Figure 5-30 if you are using the test data supplied by Microsoft for a default Azure SQL database.

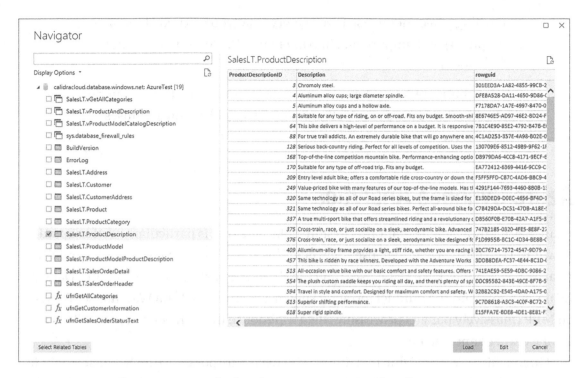

Figure 5-30. *The Navigator dialog for an Azure SQL database connection showing sample data*

Note   If you are setting up an Azure SQL database, make sure that you include firewall rules to allow connection from the computer where you are running Power BI Desktop to the Azure SQL database.

If you followed the steps to connect to an on-premises SQL Server database in Chapter 3, then you are probably feeling that the approach used here is virtually identical. Fortunately, the Power BI development team has worked hard to make the two processes as similar as possible. This extends to

- Ensuring that the DataSource settings are stored by Power BI Desktop and can be updated just as you can for an on-premises database connection

- Allowing you either to use DirectQuery (that you learned about in Chapter 4) or to import data into the Power BI Desktop in-memory data model

- Using the same Advanced options (writing your own SELECT queries or using stored procedures) that you can use with an on-premises SQL Server

Azure Synapse Analytics

Azure has many available platforms to store data. One that is particularly well adapted to Power BI Desktop is Azure Synapse Analytics. This now includes SQL Serverless pools, Apache Spark pools, Lake Databases (and the shared metastore between them), Data Explorer pools, and the product formerly known as Azure SQL Data Warehouse. This is an MPP (massively parallel processing) data warehouse that is hosted in Azure. This is what we will connect to here.

Once again, I will presume that, unless you have a corporate Azure Synapse Analytics instance at hand, you will be using a trial Azure account and that you have provisioned Azure Synapse Analytics using the sample data that Microsoft provides. Setting up a test data warehouse is very similar to preparing a database, as described at the URL at the start of the previous section. Here, too, you need firewall rules to be set up correctly (although this may not be strictly necessary if you have previously set up firewall rules for, say, Azure SQL database).

Assuming that you have access to an Azure Synapse Analytics instance:

1. Open a new Power BI Desktop application.

2. In the Power BI Desktop ribbon, click the small triangle at the bottom of the Get Data button and then click More.

3. Click Azure in the list on the left and then Azure Synapse Analytics SQL on the right.

4. Click Connect. The SQL Server database dialog will appear.

5. Enter the Azure Synapse Analytics Dedicated SQL Pool server name that you obtained from the Microsoft Azure management portal (or that was given to you by a corporate DBA). The SQL Server database dialog will look like the one shown in Figure 5-31.

This is because Azure Synapse Analytics (well, to be more accurate, the massively parallel processing system that we are connecting to here) is, essentially, a SQL Server database.

Figure 5-31. *The SQL Server database dialog for an Azure Synapse Analytics SQL connection*

6. Select the Import button and then click OK. The credentials dialog will appear.

7. Click Database on the left and enter a valid username and password.

8. Click Connect. The Navigator dialog will appear showing the database(s) that you have permission to access in the Azure Synapse Analytics Dedicated SQL Pool. This dialog will look like the one shown in Figure 5-32 if you are using the test data supplied by Microsoft.

Figure 5-32. *The Navigator dialog for an Azure Synapse Analytics connection showing sample data*

9. Select the tables that you need and click Load or Transform Data to return to Power BI Desktop and begin adding visuals to your report.

Note   Do not be phased by the fact that the title for the dialog where you specify the server and database says "SQL Server database." This will connect you to the Azure Synapse Analytics Dedicated SQL Pool correctly.

As was the case for an on-premises connection, you can choose a DirectQuery and can expand the Advanced options field to enter a specific SQL query if you are loading data.

Connecting to SQL Server on an Azure Virtual Machine

More and more databases are now hosted outside a corporate environment by cloud service providers. With a provider such as Amazon (with RDS for SQL Server) or Microsoft (who offers virtual machines—or VMs—for SQL Server in Azure), you can now site your databases outside the enterprise and access them from virtually anywhere in the world.

So, to extend the panoply of data sources available to Power BI Desktop, we will now see, briefly, how to connect to SQL Server on an Azure Virtual Machine. Admittedly, connecting to SQL Server on an Azure Virtual Machine is nearly the same as connecting to SQL Server in a corporate environment. However, it is worth a short detour to explain, briefly, how to return data to Power BI Desktop from a SQL Server instance in the cloud.

Once again, if you do not have a SQL Server instance that is hosted on an Azure Virtual Machine in your corporate environment, then you can always test this process using an Azure trial account. I cannot, however, explain here how to set up a SQL Server instance on a VM, as this is outside the scope of this book. There are, however, many resources available that can explain how to do this should you need them.

To connect to SQL Server on a Virtual Machine:

1. Open a new Power BI Desktop application.

2. In the Power BI Desktop ribbon, click the small triangle at the bottom of the Get Data button and then click SQL Server. The SQL Server database dialog will appear.

3. Enter the full string that describes the server in the Server text box. Either this will be given to you by a corporate DBA or, if you are using your own Azure account, you can find it in the Azure Management Portal.

4. Enter the database name; if you have loaded the sample data that accompanies this book into a SQL Server instance in a VM, it will be CarSalesData. The dialog will look like Figure 5-33.

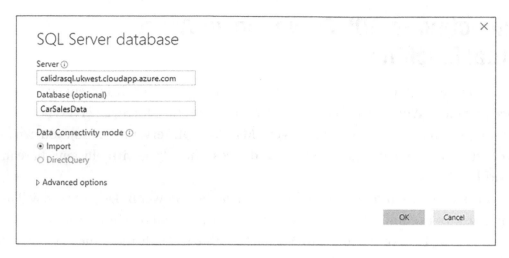

Figure 5-33. *The Microsoft SQL Server database dialog for an Azure VM*

5.   Click OK. The Access a SQL Server database dialog will appear.
     Select Database as the security mode and enter the username
     and password, as shown in Figure 5-34. If you are using your own
     Azure account, these can be the username and password that you
     specified when setting up the virtual machine.

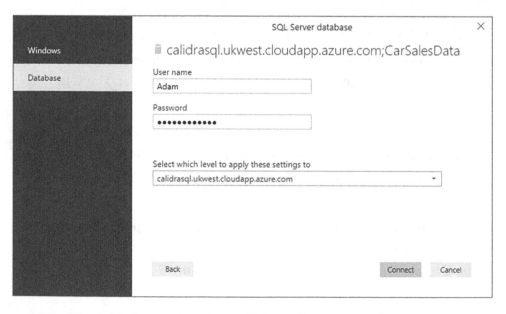

Figure 5-34. *The SQL Server database dialog when connecting to a
virtual machine*

6. If you see the encryption support dialog, click OK. The Navigator
    dialog will appear, listing all the tables that you have permissions
    to see on the SQL Server hosted by the virtual machine.

As you can see, the process is virtually identical to the one that you followed to
connect to SQL Server in Chapter 3. I have, nonetheless, a few points that I need to bring
to your attention:

- You use the Azure VM multipart name as the server name.

- As was the case when connecting to an on-premises SQL Server
  instance, you can select the database if required.

- You can use the server's IP address as the database name if the VM
  has specified a public IP address.

- Security is a big and separate question. In a corporate environment,
  you *might* be able to use Windows security to connect. You will
  almost certainly have to use database security for a test VM.

- As is always the case in Azure, firewalls must be set up correctly.

Azure Blob Storage

The final Azure data source that I want to introduce you to in this chapter is Azure Blob
Storage. So if you need to access data that is stored as files, you can connect to them via
Azure Blob Storage.

Once again, you will need either corporate access to Azure Blob Storage or an Azure
trial account. In either case, you need to copy the two sample files that are in the folder
`C:\PowerBiDesktopSamples\MultipleIdenticalFiles` into a container in your Azure
Blob Storage. Downloading the sample files is explained in Appendix A.

Once the source data is available in Azure Blob Storage, you can carry out the
following steps:

1. Open a new Power BI Desktop application.

2. In the Power BI Desktop ribbon, click the small triangle at the
    bottom of the Get Data button and then click More.

3. Click Azure in the list on the left and then Azure Blob Storage on
    the right.

4.   Click Connect. The Azure Blob Storage connection dialog will be displayed.

5.   Enter the account name that you are using to connect to Azure Blob Storage. The Azure Blob Storage dialog will look like the one shown in Figure 5-35. If you are using a corporate Azure Blob Storage account, then your system administrator will provide this. In a test scenario, you can find this in the Azure Management Portal by opening the Storage Account blade and copying the Blob Service Endpoint.

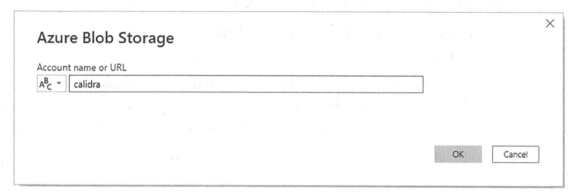

Figure 5-35.   *The Azure Blob Storage connection dialog*

6.   Click OK. The Azure Blob Storage Account key dialog will appear.

7.   In the Azure Management Portal, copy an account key. These can be found in the Azure Management Portal by clicking the Storage Account blade and then clicking Access Keys. If you have been sent an account key by a system administrator, then use that instead. Remember that account keys are highly sensitive and should be handled accordingly.

8. Paste the account key into the Azure Blob Storage Account key dialog. The dialog will look like the one in Figure 5-36.

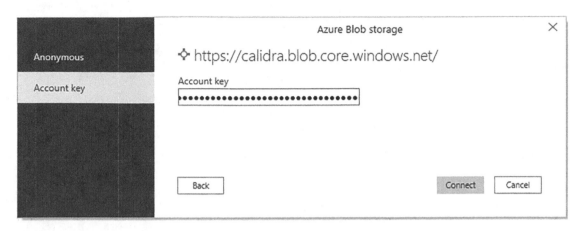

Figure 5-36. *The Azure Blob Storage Account key dialog*

9. Click Connect. The Navigator will appear, showing the list of files in the selected container. You can see an example of this in Figure 5-37.

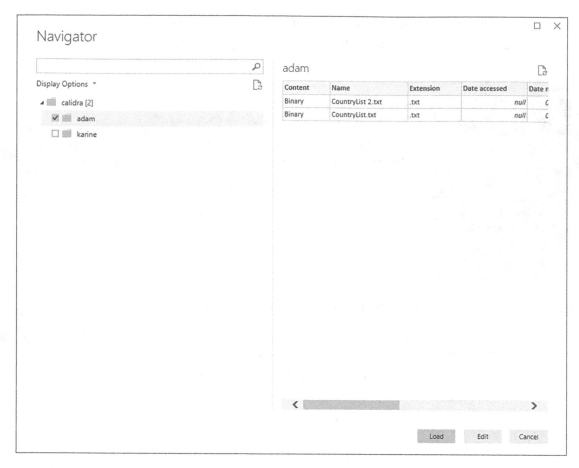

Figure 5-37. *The Navigator dialog showing available containers in Azure Blob Storage*

10. Click Load. The list of files stored in Azure will appear in Power BI Desktop.

Note   It is important to note that, for the moment at least, what you have returned from Azure is a list of available files. Chapter 10 explains how to select and load data from some or all of the available files into Power BI Desktop, where they can be used as a basis for analytics.

Azure Databricks

As a product currently in vogue (at least as this book went to press), Databricks is getting a lot of attention. So it is, perhaps, inevitable that Power BI should also be able to connect to this particular data source.

1. In the Home ribbon, click Get Data ➤ More.

2. Select Azure on the left and Azure Databricks on the right. The dialog should look like Figure 5-38.

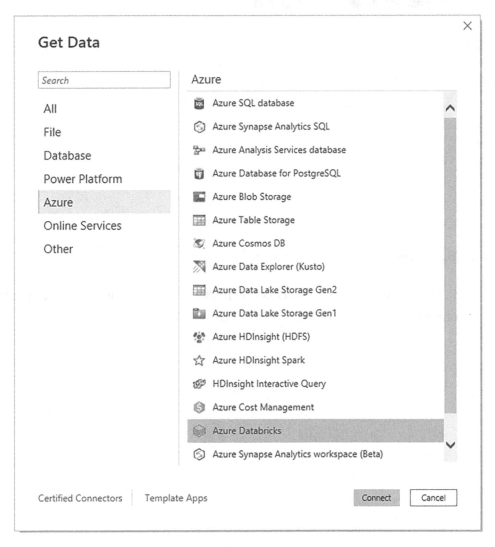

Figure 5-38. *Connecting to Databricks*

3.   Click Connect. Enter the server and HTTP path (you may need to contact your IT department for these). You should see the dialog presented in Figure 5-39.

Figure 5-39.   *Databricks connection details*

Click OK. If you have a Personal Access token, click this on the left of the Connection dialog and enter it. The dialog should look like Figure 5-40.

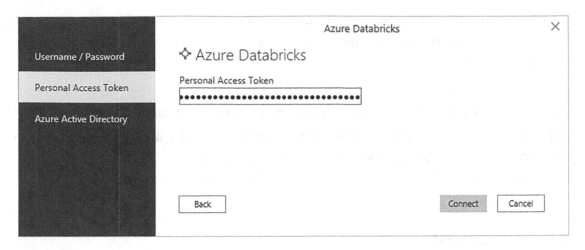

Figure 5-40. *Databricks logon using a Personal Access Token*

Click Connect. You will see the data tables that you have access rights to as shown in Figure 5-41.

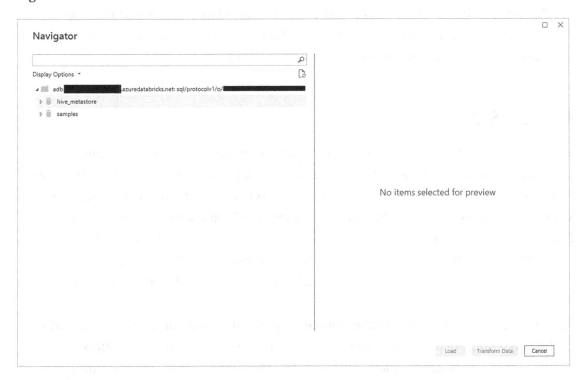

Figure 5-41. *The Navigator showing available Databricks data tables*

You may also be able to connect to Databricks using your Azure Active Directory (AD) details–or a username and password. It is best to contact your IT department for all connection credentials.

Note   As it is considered to be a database by Power BI, you can also set up DirectQuery links to Databricks data. However, you need to be aware that this can lead to extremely slow reaction times in your dashboards.

Azure Security

All cloud service providers take security extremely seriously. As you have seen in this chapter, you will always be obliged to enter some form of security token and/or specify a valid username and password to connect to cloud-based data.

All the security information that you entered is stored in the Power BI Desktop Storage Settings. This can be removed or modified in the same way that you learned to update or remove database security information in Chapter 3.

Conclusion

In this chapter, you saw an overview of how to retrieve data that you access using the Internet. This can range from a table of data on a web page to a massive Azure Synapse Analytics data repository. Alternatively, perhaps you need to create dashboards based on your Salesforce, Google Analytics, or MS Dynamics 365 data. Maybe your organization has decided to move its data centers to the cloud and is using SQL Server in Azure or Amazon Redshift. In any case, Power BI Desktop can connect and access the data available in these services and repositories. It can even access big data platforms in several cloud services.

Given the vast number of online sources, this chapter could only scratch the surface of this huge range of potential data repositories. However, as Power BI Desktop is rigorous about standardizing access to data, you should be able to apply the approaches you have learned in this chapter to many other data services, both current and future.

CHAPTER 6

Loading Data from Other Data Sources

Other Sources

There are currently many dozens of data sources for which Power BI connectors are available. Clearly, it would be impossible to explain how to use all of them. In any case, the list of available sources is growing by the month—so this would be a fruitless task. To conclude our whistle-stop tour of available source data, this chapter will introduce you to

- Power BI datasets

- Power BI dataflows

- R scripts

- Python scripts

- Dataverse

- ODBC data

- Adding your own free-form tables

- Custom connectors (which would obviously require help from a developer to set up)

If you have followed the download instructions in Appendix A, then any files used in this chapter will be in the C:\PowerBIDesktopSamples folder.

© Adam Aspin 2022
A. Aspin, *Pro Data Mashup for Power BI*, https://doi.org/10.1007/978-1-4842-8578-7_6

Power BI Datasets

Power BI datasets are complete data models that have been designed for end users and made available in the Power BI Service—a Microsoft cloud-based platform. Usually Power BI datasets have been carefully designed to include only the data and calculations that you need. They are nearly always tightly controlled by corporate data guardians.

Note   You can only access Power BI datasets if you have a Power BI license and have been given permissions by the dataset creator to access the dataset. If in doubt, consult your organization's IT department.

If you have been made aware of available Power BI datasets in your organization, you can connect to them and use them as the basis for your dashboards and reports like this:

1. Click the Power BI datasets button in the Home ribbon.

2. Sign in to the Power BI Service (unless you are already signed in). You can see the prompt to sign in in Figure 6-1.

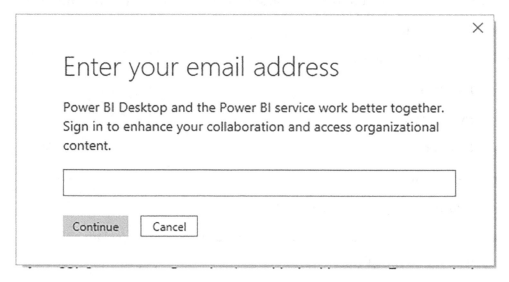

Figure 6-1. *The connection dialog for Power BI dataflows and datasets*

3. Enter your email address and once signed in, click Connect.
    You will see a list of all currently available datasets, as shown in
    Figure 6-2. Of course, you will only see the datasets that have been
    made available specifically for you.

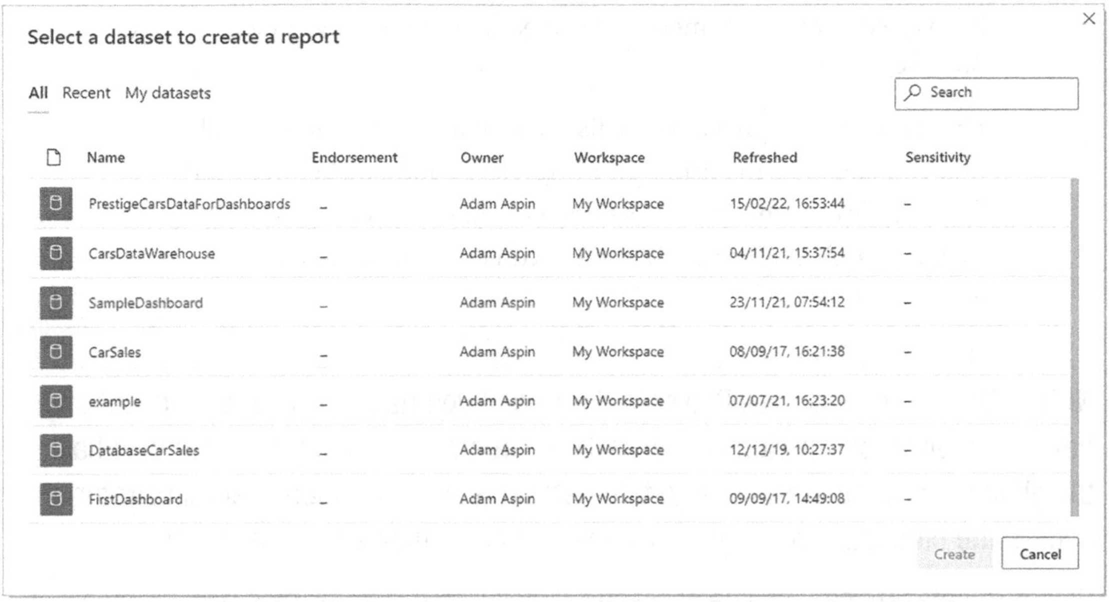

Figure 6-2. *Available datasets in the Power BI Service*

4. Select the dataset you wish to use as the basis for your dashboard
    visuals.

5. Click Create. The tables and fields that are in the dataset will
    appear in the Fields pane in Power BI Desktop.

You can now use the data in the Power BI dataset as the basis for your reports and
dashboards.

You need to be aware that once you have established the connection to a Power BI
Service Live Connection, there is very little that you can do except use the data to build
dashboards and reports (at least as this book went to press). More specifically, you
cannot transform the data using any of the techniques that are explained in Chapters 7
through 15. Nor can you add further data sources to the dataset. This may have changed
once the DirectQuery for Power BI datasets and AAS feature is released.

As it is easy to be swamped by a multitude of datasets in a corporate environment, the dataset dialog has a few options to help you to find the right dataset:

- *Recent*: Clicking this tab will only show recently created datasets.

- *My Datasets*: Clicking this tab will only show datasets that you have created or that someone else created but you have taken ownership of.

- If you are faced with a massive list of available datasets, remember that you can filter the list by entering a few characters contained in the dataset name into the search field at the top of the Datasets dialog. Simply delete the characters you entered to display the complete list.

Note   You can only access Power BI datasets if you have a Power BI account and have been given permissions by the dataset creator or by someone else who has the ability to grant permissions on that dataset (e.g., a workspace administrator) to access the dataset. If in doubt, consult your organization's IT department.

Power BI datasets can also be accessed directly from the Get data pop-up menu and in the Power Platform section of the Get Data dialog.

Power BI Dataflows

Power BI dataflows are the output of a data ingestion and transformation process in the Power BI service. They always return one or more tables of data. As is the case with Power BI datasets, Power BI dataflows are usually designed to include only the data that you will need and are nearly always tightly controlled by corporate data custodians.

Should you have been given permission to use the data in a Power BI dataflow, this is how you can connect:

1. In the Power BI Desktop Home ribbon, click the Get Data button.

2. Click Power Platform on the left, and then select Power BI dataflows on the right. The Get Data dialog will look like Figure 6-3.

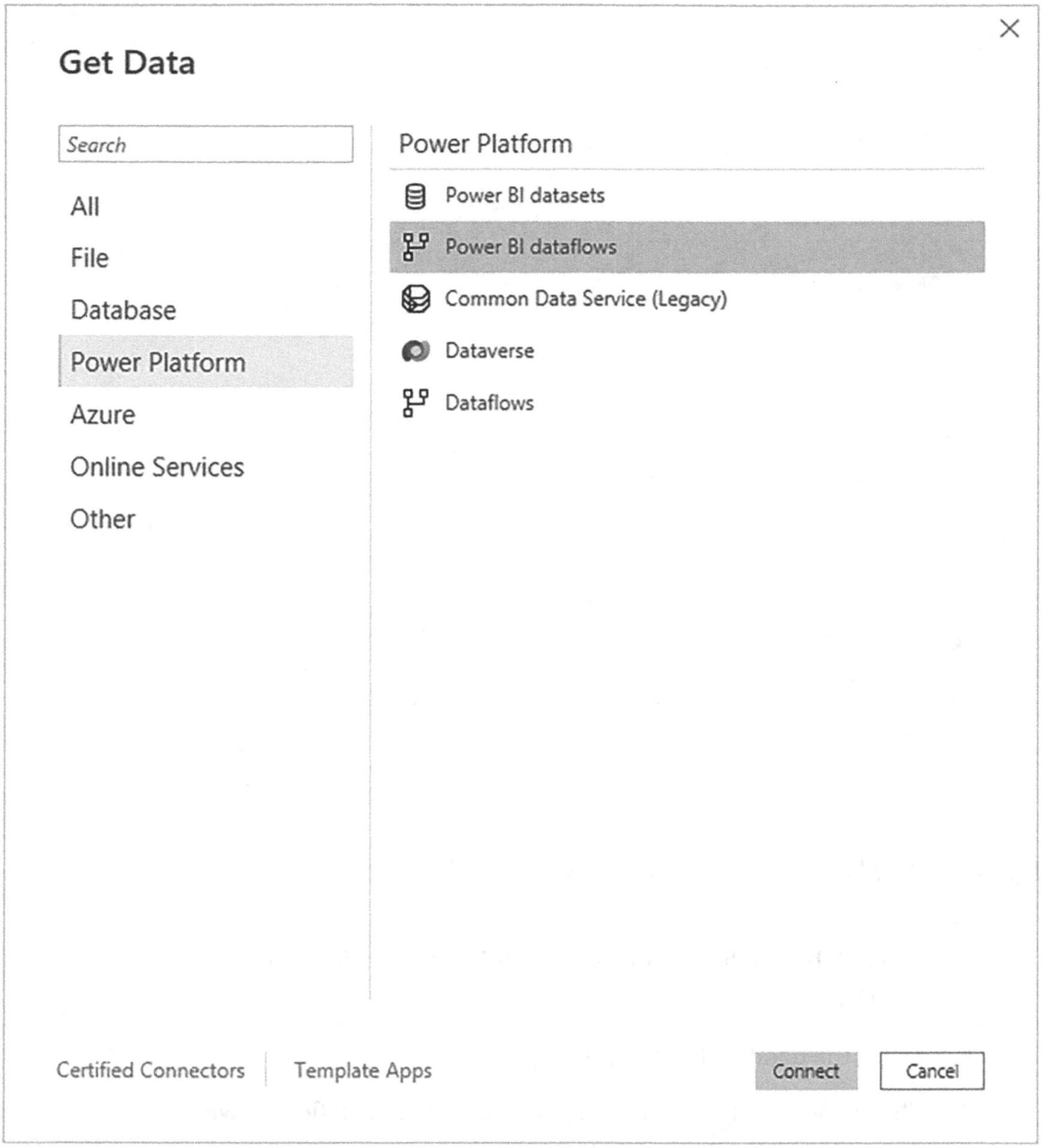

Figure 6-3. *The Get Data dialog for Power BI dataflows*

3. Sign in to the Power BI Service (unless you are already signed in).

4. Click Connect. You will see a list of all currently available dataflows in the Navigator. Of course, you will see the dataflows that have been made available specifically for you. You can see this in Figure 6-4.

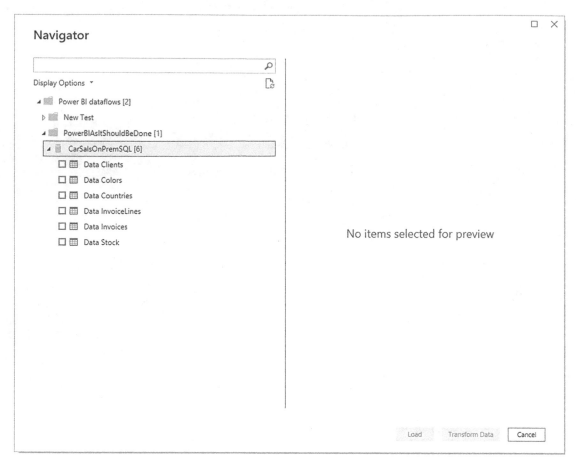

Figure 6-4. *The Get Data dialog for Power BI dataflows*

5. Expand the dataflow folder(s) containing the dataflow(s) that you want to connect to.

6. Select the dataflow(s) that you require.

7. Click Load. The tables and fields that are in the dataflow(s) will appear in the Fields pane in Power BI Desktop.

You can now use the data in the Power BI dataflow as the basis for your reports and dashboards. Unlike datasets, dataflows are simple source data and so you can modify and extend a data model based on dataflows.

Note   Power BI dataflows can also be accessed directly from the Get data pop-up menu and in the Power Platform section of the Get Data dialog.

R Scripts

Although Power BI Desktop is amazingly powerful when it comes to preparing data for dashboards and reports (and you will learn more about it in Chapters 7 through 14), there may be occasions when you prefer to use another tool to carry out some highly specific data preparation.

One such tool is the R language. R is an open source programming language for statistical analysis and data science. If you have R experience that you want to use in conjunction with Power BI, then you can try the following:

Note   To use R scripts to prepare data for loading into Power BI, you will first need to install R on your computer. One place to find R is the following URL: `https://mran.revolutionanalytics.com/download`.

1. In the Power BI Desktop Home ribbon, click the Get Data button.

2. Click Other on the left, and then select R script on the right. The Get Data dialog will look like Figure 6-5.

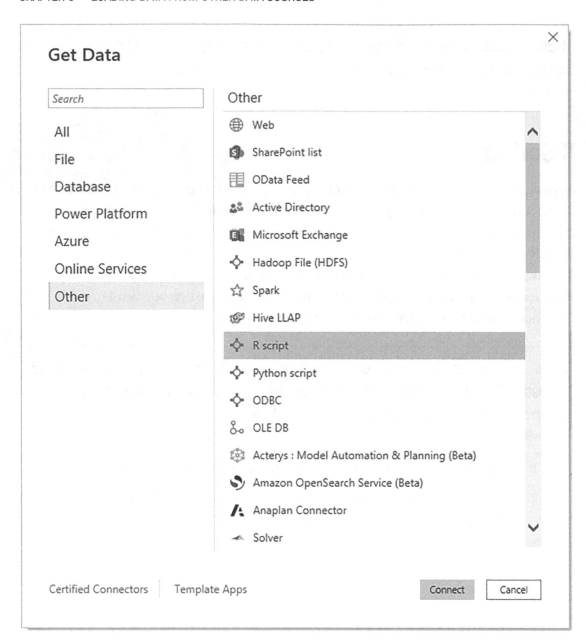

Figure 6-5. *The Get Data dialog when opting to use an R script*

3. Click Connect. The R script dialog will appear, where you can paste in the script that connects to your data and preprocesses it. If you want a sample R script, you can use the script in the file LoadR.txt from the sample files. You can see this dialog in Figure 6-6.

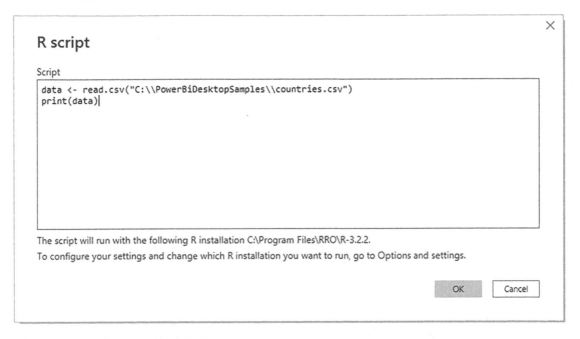

Figure 6-6. *The R script dialog*

4. Click OK. The tables and fields that are in the R script will appear in the Navigator, as shown in Figure 6-7.

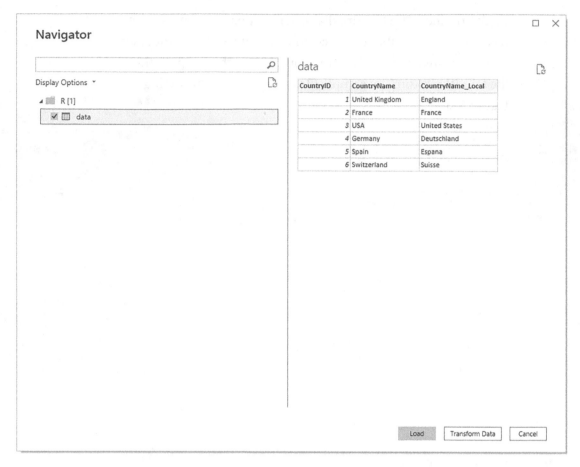

Figure 6-7. *The output from an R script in the Navigator*

5.   Click Load. The tables and fields that are in the R script will
      appear in Power BI Desktop.

This is probably the most elementary R script that you could write. Moreover, all it
does is to load a simple CSV file. Clearly, I would never suggest that you use R like this in
reality, but only use R scripts if you have them ready prepared to preprocess data sources
and provide cleansed and prepared data sets.

R is a huge subject, and I have no intention of attempting to explain even a small part
of how to use it in this book. I merely want to make R practitioners aware that they can
harness their existing knowledge and put it to the service of Power BI.

Note   The R script dialog is most emphatically *not* an R IDE (development interface). So you will probably need an R IDE to prepare and test the script that you then paste into the R script dialog.

R Options

There are a couple of R options that you can adjust if this proves really necessary. You can find these by choosing File ➤ Options and Settings ➤ Options (or, more simply, by clicking the options icon at the top right of the R script editor) and then clicking R scripting in the left-hand pane of the Options dialog that you can see in Figure 6-8.

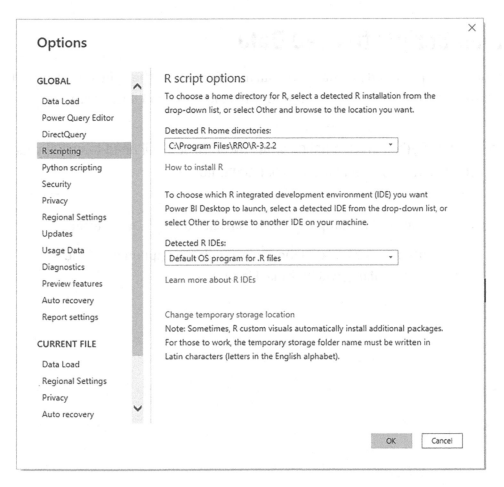

Figure 6-8.   *R options*

There are, essentially, only a couple of possible tweaks that you could have to make:

- *R scripting engine*: Power BI Desktop should detect the R scripting engine that you have previously installed. Should it not do this, then you can browse to the directory containing the R scripting engine. Simply select Other in the pop-up list and click the Browse button.

- *IDE*: If you prefer to use a different external R integrated development environment (IDE), then simply click the Browse button to select a preferred R IDE. This will then be invoked the next time that you click the External Editor icon at the top right of the R script editor. This is only relevant for the "R script visual," and not R scripts in Power Query.

Python Scripts to Load Data

If Python is your preferred language for data transformation, then you can also apply your existing Python knowledge to preparing data for Power BI dashboards.

Note   To use Python in conjunction with Power BI, you will need to install your preferred Python version and IDE on your computer.

---

1. In the Power BI Desktop Home ribbon, click the Get Data button.

2. Click Other on the left, and then select Python script on the right. The Get Data dialog will look like Figure 6-9.

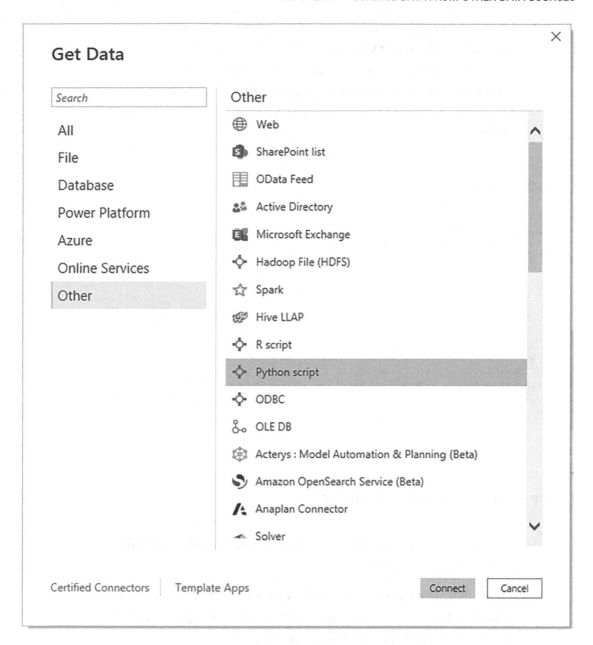

Figure 6-9. *The Get Data dialog when opting to use a Python script*

3. Click Connect. The Python script dialog will appear, where you
    can paste in the script you previously prepared in a Python
    IDE that connects to your data and preprocesses it. If you
    want a sample Python script, you can use the script in the file
    LoadPython.txt from the sample files. You can see this dialog in
    Figure 6-10.

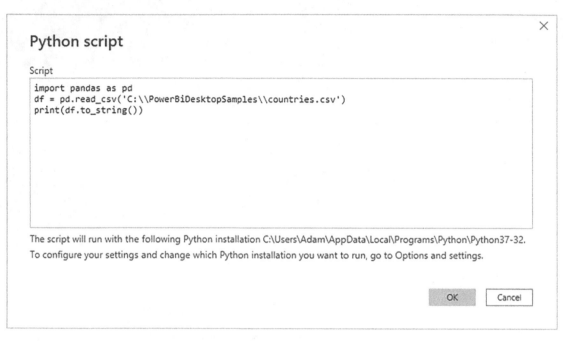

Figure 6-10. *The Python script dialog*

4. Click OK. The tables and fields that are in the Python script will
    appear in the Navigator.

5. Click Load. The tables and fields that are in the Python script will
    appear in Power BI Desktop, as shown in Figure 6-7 (as both the R
    and Python scripts load the same source file).

Note   I have no intention of attempting to explain how to use Python in this book.
I merely want to make Python practitioners aware that they can harness their
existing knowledge and put it to the service of Power BI.

Python Options

As was the case with R visuals, you can tweak a couple of options to specify Python script options for Power BI Desktop. As these are essentially identical to those described previously for R, I will not repeat these elements here. You can, however, see the available options in Figure 6-11.

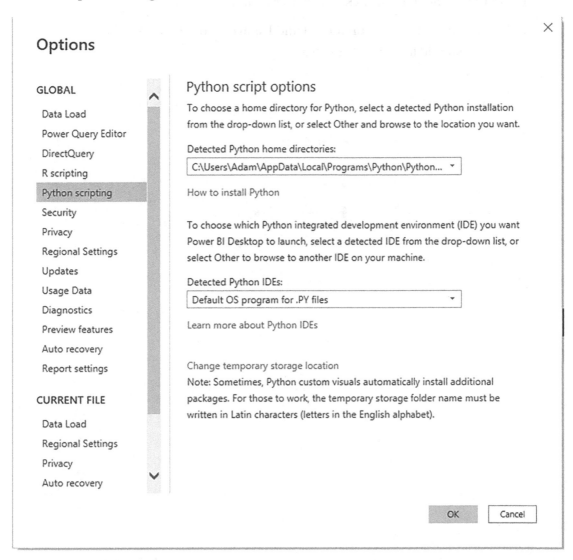

Figure 6-11. *Python options*

Dataverse

If you are using data stored in the Microsoft Dataverse (which is part of what used to be called the Common Data Service), then you can connect to this extremely easily—as you would probably expect as the Dataverse is a Microsoft product.

1. In the Home ribbon, click Get Data ➤ More.

2. Select Power Platform on the left and Dataverse on the right. The dialog should look like Figure 6-12.

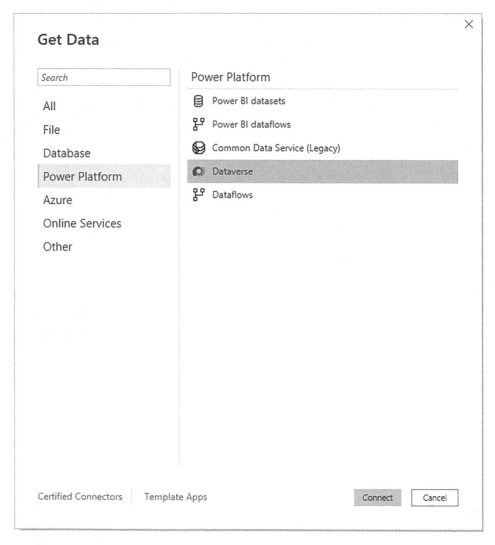

Figure 6-12. *Connecting to the Common Data Service*

3. Click Connect. You will be prompted to log in as shown in Figure 6-13.

Figure 6-13. *Connecting to the Dataverse*

4. Click Connect. The Navigator will appear showing all accessible data objects in the Dataverse.

5. Click OK. The Navigator will appear where you can select the data entities that you wish to connect to.

As Dataverse is essentially a huge database, I suggest that you refer back to Chapter 3 for details on using database sources for Power BI Desktop.

ODBC Sources

As you have seen in this chapter and the preceding one, Power BI Desktop can connect to a wide range of data sources. However, there will always be database applications for which there is no specific connector built into Power BI Desktop.

This is where a generic solution called Open Database Connectivity (or ODBC) comes into play. ODBC is a standard way to connect to data sources, most of which are databases or structured like databases. Simply put, if an ODBC driver exists for the application that you want to connect to, then you can load data from it into Power BI Desktop.

Hundreds of ODBC drivers have been written. Some are freely available; others require you to purchase a license. They exist for a wide spectrum of applications ranging from those found on most PCs to niche products.

Although ODBC is designed as a standard way of accessing data in applications, each ODBC driver is slightly different from every other ODBC driver. Consequently, you might have to spend a little time learning the quirks of the interface for the driver that comes with the application that you want to connect to.

In this section, we will use FileMaker Pro as a data source. This product is a desktop and server database system that has been around for quite some time. However, there is currently no specific Power BI Desktop connector for it. The good news is that FileMaker Pro *does* have an ODBC driver. So we will use ODBC to connect to FileMaker Pro from Power BI Desktop.

I have to add that I am not expecting you to install a copy (even if it is only a trial copy) of FileMaker Pro and its companion ODBC driver to carry out this exercise. What I do want to explain, however, is how you can use ODBC to connect to a wide range of data sources where an ODBC driver is available. So feel free to download and install FileMaker Pro and its ODBC driver if you wish, but you will have to refer to the FileMaker Pro documentation for an explanation of how to do this.

Assuming that you have an ODBC-compliant data source and a working ODBC driver for this data source, here is how to load data into Power BI Desktop using ODBC:

1. Run the ODBC Data Source Administrator app. This is normally in the folder `C:\ProgramData\Microsoft\Windows\Start Menu\ Programs\Administrative` Tools. Be sure to use the 64-bit version if you are using 64-bit Power BI Desktop or the 32-bit version if you are using 32-bit Power BI Desktop.

2. Click the System DSN tab. You should see the dialog shown in Figure 6-14.

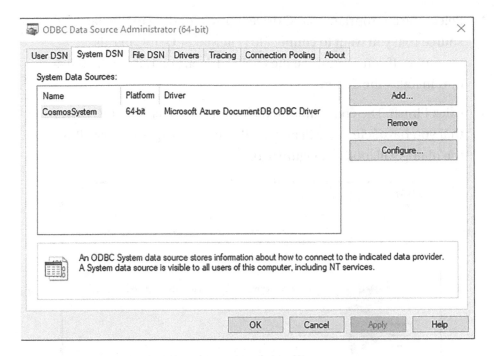

Figure 6-14. *The ODBC Data Source Administrator*

3. Click Add. You will see the list of all currently installed ODBC drivers on your computer. This should look something like the dialog shown in Figure 6-15.

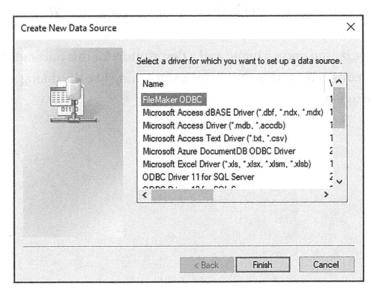

Figure 6-15. *The list of installed ODBC drivers*

4. Select the appropriate ODBC driver corresponding to the data source that you want to connect to (FileMaker ODBC in this example). If you cannot see the ODBC driver, you need to install— or reinstall—the driver.

5. Click Finish. The configuration dialog for the specific ODBC driver that you have selected will appear. If you are using FileMaker Pro, the dialog will look like Figure 6-16.

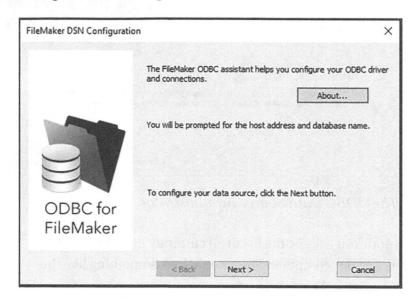

Figure 6-16. *The FileMaker Pro ODBC configuration assistant*

6. Click Next, and enter a name and a description for this particular ODBC connection. This should look something like the dialog shown in Figure 6-17.

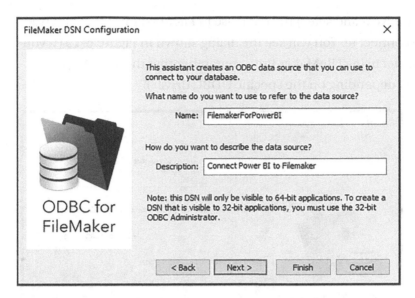

Figure 6-17. *Naming the ODBC connection for FileMaker Pro*

7. Click Next and enter **localhost** as the hostname if you are using a FileMaker trial version on your local computer. Otherwise, enter the IP address of the FileMaker server. You should see the dialog shown in Figure 6-18.

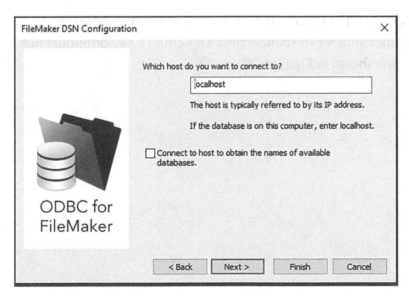

Figure 6-18. *Specifying the host for the ODBC data*

8. Click Next and select the database in FileMaker Pro that you want to connect to. You will see the dialog shown in Figure 6-19 (if you are *not* using FileMaker Pro—remember that these dialogs can vary depending on the specific ODBC driver).

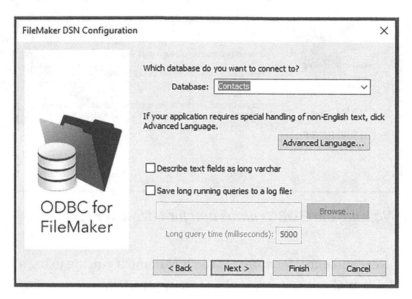

Figure 6-19. *Specifying the database for the ODBC data*

9. Click Next. The ODBC configuration dialog will resume the specifications for the connection. This could look something like the one shown in Figure 6-20.

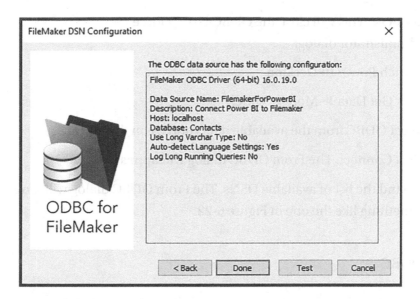

Figure 6-20. *The ODBC connection confirmation dialog*

10. Click Done. You will return to the ODBC Data Source
     Administrator, where you will see the System DSN that you just
     created. The ODBC Data Source Administrator dialog should look
     something like the one shown in Figure 6-21.

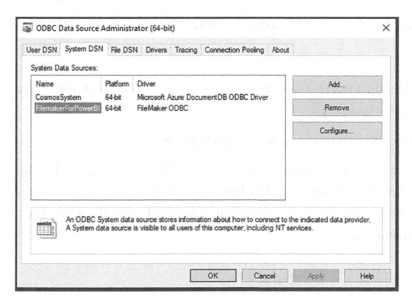

Figure 6-21. *The ODBC Data Source Administrator dialog with an ODBC driver configured*

11. Click OK. This will close the ODBC Data Source Administrator dialog.

12. Launch Power BI Desktop.

13. Click Get Data ➤ More ➤ Other.

14. Select ODBC from the available data sources on the right.

15. Click Connect. The From ODBC dialog will appear.

16. Expand the list of available DSNs. The From ODBC dialog will look something like the one in Figure 6-22.

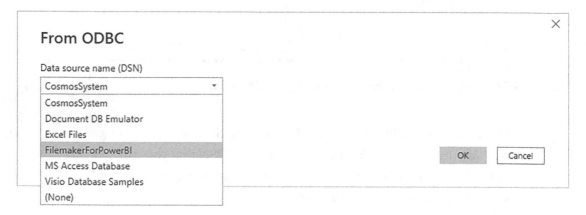

Figure 6-22. *The Power BI Desktop From ODBC dialog to select an ODBC data source*

17. Select the DSN that you created previously (FilemakerForPowerBI in this example).

18. Click OK. The credentials dialog will appear.

19. Choose Windows integrated security or click Database on the left and enter the username that has permissions to connect using the ODBC driver. The credentials dialog will look something like the one in Figure 6-23.

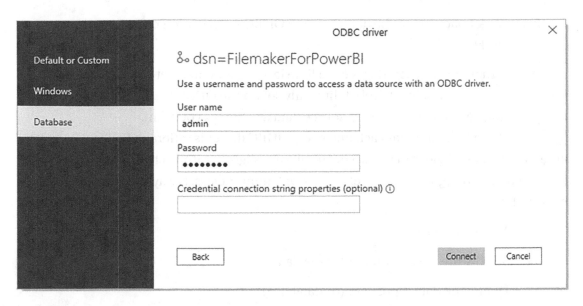

Figure 6-23. *The ODBC driver security dialog*

20.   Click Connect. You will see the data that is available in the ODBC data source in the Navigator window.

21.   Select the table(s) that you want to load into Power BI Desktop. You can see the data contained in the selected table in Figure 6-24.

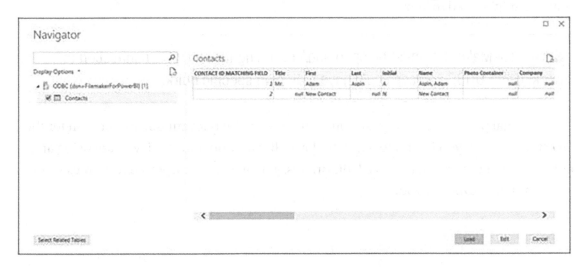

Figure 6-24. *The Navigator dialog when using an ODBC source dialog*

22. Click Load to load the data from the ODBC source into Power BI
Desktop.

I realize that this process may seem a little laborious at first. Yet you have to
remember that you will, in all probability, only set up the ODBC connection once. After
that, you can use it to connect to the source data as often as you want.

You need to be aware that each and every ODBC driver is different. So the
appearance of the dialogs in steps 5 to 10 will vary slightly with each different ODBC
driver that you configure. The key elements will, nonetheless, always be the same. They
are as follows:

- Name the DSN.

- Specify the host computer for the data.

- Define the data repository (or database).

There is much more that could be written about creating and using ODBC
connections to load data into Power BI Desktop—or indeed into any number of
destination applications. However, I will have to refer you to the wealth of available
resources both in print and online if you need to learn more about this particular
technology. A good starting point is the Microsoft documentation that explains the
difference between system, user, and file DSNs and describes many of the key elements
that you might need to know.

Note   FileMaker Pro must be open and/or running for an ODBC connection to
work. Other ODBC sources could have their own specific quirks.

As a final point, I can only urge you to procure all the relevant documentation for the
ODBC driver that you intend to use with Power BI Desktop. Indeed, if you are using an
enterprise data source that uses ODBC drivers, you may have corporate resources who
can configure ODBC for you.

Refreshing Data

Loading data from databases and data warehouses only means that a snapshot of the source data is copied into Power BI Desktop. If the source data is updated, extended, or deleted, then you will need to get the latest version of the data (unless you're using DirectQuery) if you want your analyses to reflect the current state of the data.

Essentially you have two options to do this:

- Refresh all the source data from all the data sources that you have defined.

- Refresh one or more tables individually.

Refreshing the Entire Data in the Power BI Desktop In-Memory Model

There is only one way to be certain that all your data is up to date. Refreshing the entire data may take longer, but you will be sure that your Power BI Desktop file contains the latest available data from all the sources that you have connected to.

To carry out a complete refresh:

1. In the Home ribbon, click the Refresh button. The Refresh dialog will appear, showing all the data sources that are currently being refreshed. This dialog will look like the one in Figure 6-25.

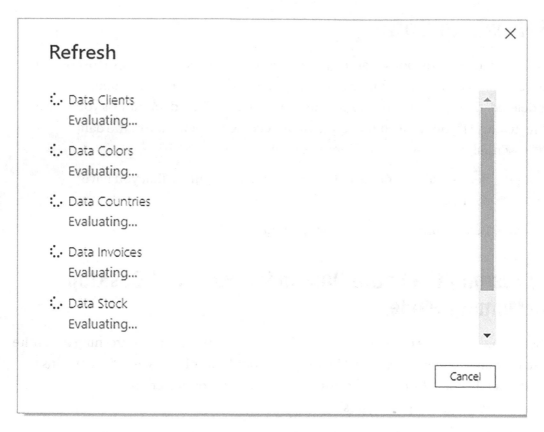

Figure 6-25. *The Refresh dialog*

Note   A full data refresh can take quite a while if the source data is voluminous or
if the network connection is slow.

Refreshing an Individual Table

If you are certain that only one or more tables need to be refreshed in your Power BI
Desktop data model, then you can choose to refresh tables individually. To do this:

1.   In the Fields pane, right-click the table that you want to refresh.

2.   Select Refresh data in the context menu. This is illustrated in
      Figure 6-26.

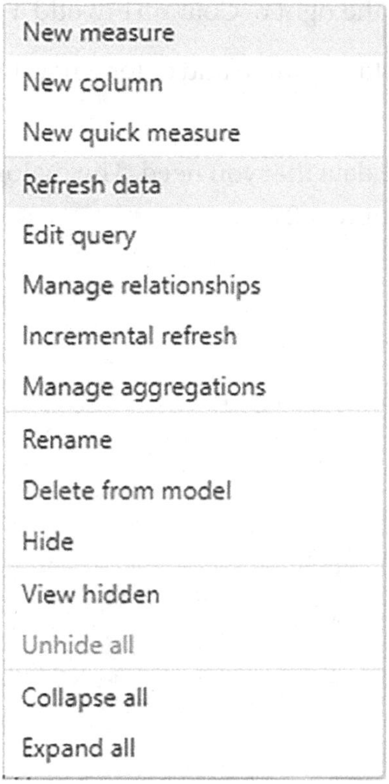

New measure
New column
New quick measure
Refresh data
Edit query
Manage relationships
Incremental refresh
Manage aggregations
Rename
Delete from model
Hide
View hidden
Unhide all
Collapse all
Expand all

Figure 6-26. *Refreshing a single table*

The Refresh dialog will appear (possibly only briefly), and the existing data for this table will be replaced with the latest data.

Adding Your Own Data

All the data you need may not be always available. You might find yourself needing to add a list of products, a group of people, or, indeed, any kind of data to the datasets that you have loaded into Power BI Desktop.

The development team at Microsoft has recognized this need and offers a simple solution: you can create your own tables of data to complete the collection of datasets in a Power BI in-memory data model. Then you can enter any extra data that you need, on the fly.

1. In the Power BI Desktop Home ribbon, click Enter Data. The Create Table dialog will appear.

2. Click the asterisk to the right of Column1 to add a column.

3. Double-click any column name and enter a new name to rename the column.

4. Enter or paste in the data that you need. The dialog will look like the one shown in Figure 6-27.

Figure 6-27. *The Create Table dialog*

5. Enter a name for the table in the Name field at the bottom of the dialog.

6. Click Load to load the data into the Power BI in-memory data model. The newly created table will appear in the Fields pane.

Editing facilities in the Create Table dialog are extremely simplistic. You can delete, cut, copy, and paste data and columns, but that is about all that you can do. However, this option can, nonetheless, be extremely useful when you need to add some last-minute data to a model.

Note   This is an extremely simple process that is designed for small amounts of data. If you need more than just a handful of rows and columns, you could be better served by creating the data in Excel and then loading it into Power BI Desktop.

Despite the simplicity, there are a few things that you need to know to get the most out of this option:

- Right-clicking any row or column in the table grid displays a context menu where you can choose to insert or delete rows and columns.

- The Enter Data option is also available in the Query Editor.

- Once created, the table can be modified in the Query Editor just like any other data source.

- Once created, the data is "hard-coded" into Power BI Desktop.

Conclusion

In this chapter, you have seen how to connect to a range of data sources—both in Azure and on your computer or network. You have seen that your organization's existing Power BI Service can contain pre-prepared dataflows and datasets that you can connect to. You then learned that you can capitalize on your R and Python knowledge and use these two languages to prepare data for use by Power BI. You have also learned that you can use the Dataverse as the basis for your Power BI reports and dashboards.

Finally, you saw that you can use one of the standard data access connectors— ODBC—to connect to sources of data for which there is not (yet) a built-in Power BI connector. Then, if all else fails, you can create your own table of data should you need to.

This chapter concludes the set of six chapters that introduced you to some of the many and varied data sources that you can use with Power BI Desktop. In the course of these chapters, you have seen how to load data from a selection of the more frequently used available sources. The good news is that Power BI Desktop can read data from dozens more sources. The bad news is that it would take a whole book to go into all of them in detail.

So I will not be describing any other data sources in this book. This is because now that you have come to appreciate the core techniques that make up the extremely standardized approach that Power BI Desktop takes to loading data, you can probably load any possible data type without needing much more information from me. Should you need any specific information on other data sources, then your best port of call is the Microsoft Power BI website. This is currently at `https://powerbi.microsoft.com/`.

Now that you can find, access, and load the data you need into Power BI Desktop, it is time to move on to the next step. This means cleansing and restructuring the datasets so that they suit your analytical requirements. Handling these challenges is the subject of the remainder of this book.

CHAPTER 7

Power Query

In the previous six chapters, you saw some of the ways in which you can find and load (or connect to) data for the Power BI Desktop data model to use. Inevitably, this is the first part of any process that you follow to extract, transform, and load data. Yet it is quite definitely only a first step. Once the data is in Power BI Desktop, you need to know how to adapt it to suit your requirements in a multitude of ways. This is because not all data is ready to be used immediately. Quite often, you have to do some initial work on the data to make it usable in Power BI Desktop. Tweaking source data is generally referred to as *data transformation*, which is the subject of this chapter as well as the next six.

The range of transformations that Power BI Desktop offers is extensive and varied. Learning to apply the techniques that Power BI Desktop makes available enables you to take data as you find it, then cleanse it, and push it back, cleansed and prepared, into the Power BI Desktop data model as a series of coherent and structured data tables. Only then is it ready to be used to create compelling dashboards and reports.

As it is all too easy to be overwhelmed (at least initially) by the extent of the data transformation options that Power BI Desktop has to offer, I have grouped the possible modifications into four categories. These categories are my own and are merely a suggestion to facilitate understanding:

- *Data transformation*: This includes removing columns and rows, renaming columns, as well as filtering data.

- *Data modification*: This covers altering the actual data in the rows and columns of a dataset.

- *Extending datasets*: This encompasses adding further columns, possibly expanding existing columns into more columns or rows, and adding calculations.

- *Joining datasets*: This involves combining multiple separate datasets—possibly from different data sources—into a single dataset.

© Adam Aspin 2022
A. Aspin, *Pro Data Mashup for Power BI*, https://doi.org/10.1007/978-1-4842-8578-7_7

All of this can be performed using the key tool that you will use to transform data. It is called Power Query–and is a fundamental part of Power BI Desktop.

Power Query, however, is not immediately visible when you launch Power BI Desktop. Indeed, it is probably best to consider it as part of the Power BI infrastructure–essential but hidden, yet always accessible when you need it.

As data wrangling is very different to creating dashboards, the Power Query interface is very different to the interface used for assembling and formatting visuals. This chapter will introduce you to the Power Query interface as the prerequisite for learning how to transform data.

In the following chapters, you will build on this basic knowledge and learn how to restructure data tables and cleanse and modify data.

In this chapter, I will also use a set of example files that you can find on the Apress website. If you have followed the instructions in Appendix A, then these files are in the C:\PowerBiDesktopSamples folder.

Power BI Desktop Queries

In Chapter 1, you saw how to load source data directly into Power BI Desktop from where you can use it immediately to create dashboards. Clearly, this approach presumes that the data that you are using is perfectly structured, clean, and error-free. Source data is nearly always correct and ready to use in reports and dashboards when it comes from "corporate" data sources such as data warehouses (held in relational, dimensional, or tabular databases). This is not always the case when you are faced with multiple disparate sources of data that have not been precleansed and prepared by an IT department. The everyday reality is that you could have to cleanse and transform much of the source data that you will use for your Power BI Desktop dashboards.

The really good news is that the kind of data transformation that used to require expensive servers and industrial-strength software is now available for free. Yes, Power BI Desktop comes with an awesome ETL (Extract, Transform, and Load) tool that can rival many applications that cost hundreds of thousands of dollars.

Power BI Desktop data transformation is carried out using *queries*. As you saw in previous chapters, you do not have to modify source data. You can load it directly if it is ready for use. Yet if you need to cleanse the data, you add an intermediate step between connecting the data and loading it into the Power BI Desktop data model. This intermediate step uses Power Query to tweak the source data.

So how do you apply queries to transform your data? You have two choices:

- Load the data first from one or more sources, and then transform it later.

- Edit each source data element in a query before loading it.

Power BI Desktop is extremely forgiving. It does not force you to select one or the other method and then lock you into the consequences of your decision. You can load data first and then realize that it needs some adjustment, switch to Power Query and make changes, and then return to creating your dashboard. Or you can first focus on the data and try to get it as polished and perfect as possible before you start building reports. The choice is entirely up to you.

To make this point, let's take a look at both of these ways of working.

Note   At risk of seeming pedantic and old-fashioned, I would advise you to make notes when creating really complex transformations, because going back to a solution and trying to make adjustments later can be painful when they are not documented at all.

Editing Data After a Data Load

In Chapter 1, you saw how to load the Excel workbook CarSales.xlsx directly into the Power BI Desktop data model to use it to create a starter dashboard. Now let's presume that you want to make some changes to the structure of the data that you have already loaded. Specifically, you want to rename the CostPrice column. The file that you want to modify is Example1.pbix file in the C:\PowerBiDesktopSamples directory.

1. Launch Power BI Desktop.

2. Open the sample file C:\PowerBiDesktopSamples\Example1.
   pbix. Note that in this example, you are taking an existing Power BI Desktop file and modifying the existing data input process and *not* ingesting data. Take a look at the Fields list (you may need to expand the BaseData table to do this) and note that there is a field named CostPrice.

3. In the Power BI Desktop Home ribbon, click the Transform Data button. Power Query will open and display the source data as a table. The window will look like Figure 7-1.

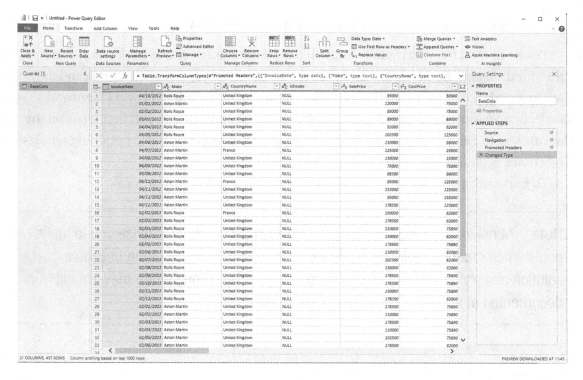

Figure 7-1. *Power Query*

4. Right-click the title of the CostPrice column (you may need to scroll right through the columns of data to find it). The column will be selected and the title will appear in yellow.

5. Select Rename from the context menu. You can see the context menu in Figure 7-2.

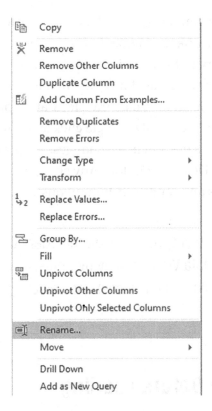

Figure 7-2. *The column context menu in the Query Editor*

6. Type **VehicleCost** and press Enter. The column title will change to VehicleCost. A new step–called Renamed Columns–will appear at the bottom of the Applied Steps list on the right of the Power Query window.

7. In the Power Query Home ribbon, click the Close & Apply button. Power Query will close and return you to the Power BI Desktop window. VehicleCost has replaced CostPrice anywhere that it was used in the dashboard. This is visible in the Fields list and all visuals that used this field.

Tip   Another simple way to rename a column in Power Query is to double-click the column name.

I hope that this simple example makes it clear that transforming the source data is a quick and painless process. The technique that you applied—renaming a column—is only one of many dozens of possible techniques that you can apply to transform your data. However, it is not the specific transformation that is the core idea to take away here. What you need to remember is that the data transformation process that underpins the data feeding into your dashboard is only a single click away. At any time, you can "flip" to the data transformation process and make changes, simply by clicking the Transform Data button in the Power BI Desktop window. Any changes that you make and confirm will update your dashboards and reports instantaneously.

Note   Alternatively, if you want, you can edit the query behind any table that is visible in the Report or Data Views simply by right-clicking the table name in the Queries pane (or clicking the ellipses) and selecting Edit Query from the context menu.

Transforming Data Before Loading

On some occasions, you might prefer to juggle with your data before you load it. This is a variation on the approach that you have used in Chapter 1 when creating a simple dashboard. Do the following to transform the data you are ingesting *before* it appears in the Power BI Desktop window:

1. Open a new Power BI Desktop window and close the splash screen.

2. In the Power BI Desktop Home ribbon, click the tiny triangle in the Get Data button.

3. Select Excel in the menu and open the Excel file
   `C:\PowerBiDesktopSamples\CarSales.xlsx`.

4. In the Navigator window, select the BaseData worksheet.

5. Click the Transform Data button (*not* the Load button).

6. Power Query will open and display the source data as a table.

7. Carry out steps 4 through 6 from the previous example to rename the CostPrice column.

8. In the Power Query Home ribbon, click the Close & Apply button. Power Query will close and return you to the Power BI Desktop window. You will see the Apply query changes dialog while the data is loaded, like the one that you can see in Figure 7-3.

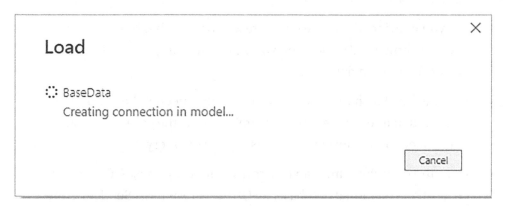

Figure 7-3. *The Apply query changes dialog*

This time, you have made a simple modification to the data *before* loading the dataset into the Power BI Desktop data model. The data modification technique was exactly the same. The only difference between loading the data directly and taking a detour via Power Query was clicking Edit instead of Load in the Navigator dialog. This means that the data was only loaded once you had finished making any modifications to the source data in Power Query.

Tip   What is also worth remembering is that these changes have no effect on the source data, which remains untouched. Modifications only happen to the copy of the data in Power BI Desktop.

Transform or Load?

Power BI Desktop always gives you the choice of loading data directly into its data model or taking a constructive detour via Power Query. The path that you follow is entirely up to you and clearly depends on each set of circumstances. Nonetheless, you might want to consider the following basic principles when faced with a new dashboarding challenge using unfamiliar data:

- Are you convinced that the data is ready to use? That is, is it clean and well structured? If so, then you can try loading it directly into the Power BI Desktop data model.

- Are you faced with multiple data sources that need to be combined and molded into a coherent structure? If this is the case, then you really need to transform the data using Power Query.

- Does the data come from an enterprise data warehouse? This could be held in a relational database, a SQL Server Analysis Services cube, or even an in-memory tabular data warehouse. Alternatively, is the data in an online service? As these data sources are nearly always the result of many hundreds—or even thousands—of hours of work cleansing, preparing, and structuring the data, you can probably load these straight into the data model.

- Does the data need to be preaggregated and filtered? Think Power Query.

- Are you likely to need to change the field names to make the data more manageable? It could be simpler to load the data directly into the data model and change them there.

- Are you faced with a lot of lookup tables that need to be added to a "core" data table? Then Power Query is your friend.

- Does the data contain many superfluous or erroneous elements? Then use Power Query to remove these as a first step.

- Does the data need to be rationalized and standardized to make it easier to handle? In this case, the path to success is via Power Query.

- Is the data source enormous? If this is the case, you could save time by editing and filtering the data first in the Query Editor. This is because the Query Editor only loads a *sample* of the data for you to tweak. The entire dataset will only be loaded when you confirm all your modifications and close the Query Editor.

These kinds of questions are only rough guidelines. Yet they can help to point you in the right direction when you are working with Power BI Desktop. Inevitably, the more that you work with this application, the more you will develop the reflexes and intuition that will help you make the correct decisions. Remember, however, that Power BI Desktop is there to help and that even a directly loaded dataset is based on a query. So you can always load data and then decide to tweak the query structure later if you need to. Alternatively, editing data in a Power Query can be a great opportunity to take a closer look at your data before loading it into the data model—and it only adds a couple of clicks. Indeed, defaulting to "Transform Data" is often–but not always–the better option. At the very least, it can help you to get a better feel of the data that you are working with.

So feel free to adopt a way of working that you feel happy with. Power BI Desktop will adapt to your style easily and almost invisibly, letting you switch from data to dashboards so fluidly that it will likely become second nature.

The upcoming chapters will take you through many of the core techniques that you need to know to cleanse and shape your data. However, before getting into all the detail, let's spend the rest of this chapter on a quick, high-level look at Power Query and the way that it is laid out.

Power Query

All of your data transformation will take place in Power Query. It is a separate window from the one where you create your dashboards, and it has a slightly different layout.

Power Query consists of six main elements:

- The six main ribbons: Home, Transform, Add Column, View, Tools, and Help. Other ribbons are available when carrying out specific types of data transformations.

- The Query list pane containing all the queries that have been added to a Power BI Desktop file.

- The Data window, where you can see a sample of the data for a selected query.

- The Query Settings pane that contains the list of steps used to transform data as well as metadata/properties about the query as a whole (name, description, enable load, include in report refresh).

- The formula bar above the data that shows the code (written in the Power BI "M" language that you will see in greater detail in Chapter 14) that performs the selected transformation step.

- The status bar (at the bottom of the window) that indicates useful information, such as the number of rows and columns in a query table, and the date when the dataset was downloaded.

The callouts for these elements are shown in Figure 7-4.

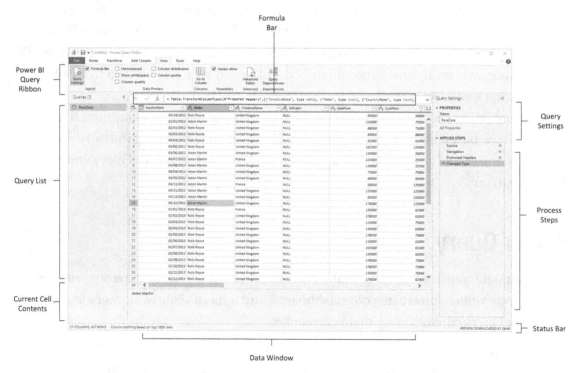

Figure 7-4. *Power Query, explained*

The Applied Steps List

Data transformation is by its very nature a sequential process. So the Query window stores each modification that you make when you are cleansing and shaping source data. The various elements that make up a data transformation process are listed in the Applied Steps list of the Query Settings pane in the Query Editor.

Power Query does not number the steps in a data transformation process, but it certainly remembers each one. They start at the top of the Applied Steps list (nearly always with the Source step) and can extend to dozens of individual steps that trace the evolution of your data until you load it into the data model. You can, if you want, consider the Query Editor as a kind of "macro recorder."

Moreover, as you click each step in the Applied Steps list, the data in the Data window changes to reflect the results of each transformation, giving you a complete and visible trail of all the modifications that you have applied to the dataset.

The Applied Steps list gives a distinct name to the step for each and every data modification option that you cover in this chapter and the next. As it can be important to understand exactly what each function actually achieves, I will always draw to your attention the standard name that Power Query applies.

Power Query Ribbons

Power Query uses (in the April 2022 version, at least) six core ribbons. They are fundamental to what you learn in the course of this chapter. They are as follows:

- The Home ribbon

- The Transform ribbon

- The Add Column ribbon

- The View ribbon

- The Tools ribbon

- The Help ribbon

I am not suggesting for a second that you need to memorize what all the buttons in these ribbons do. What I hope is that you are able to use the following brief descriptions of the Query Editor ribbon buttons to get an idea of the amazing power of Power BI Desktop in the field of data transformation. So if you have an initial dataset that is not quite as you need it, you can take a look at the resources that Power BI Desktop has to offer and how they can help. Once you find the function that does what you are looking for, you can jump to the relevant section for the full details on how to apply it.

The Home Ribbon

Since we will be making intense use of the Power Query Home ribbon to transform data, it is important to have an idea of what it can do. I explain the various options in Figure 7-5 and in Table 7-1.

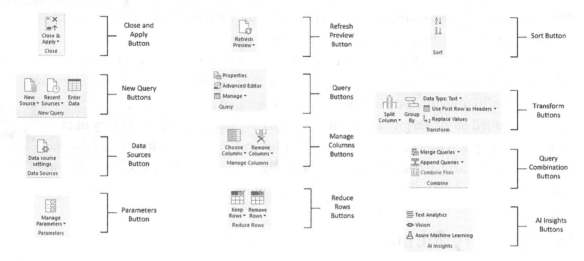

Figure 7-5. *The Query Editor Home ribbon*

Table 7-1. *Query Editor Home Ribbon Options*

Option	Description
Close & Apply	Finishes the processing steps; saves and closes the query.
New Source	Lets you discover and add a new data source to the existing query set.
Recent Sources	Lists all the recent data sources that you have used.
Enter Data	Lets you add your own specific data in a custom table.
Data Source Settings	Allows you to manage and edit settings for data sources that you have already connected to.
Manage Parameters	Lets you view and modify any parameters defined for this Power BI Desktop file. These are explained in Chapter 13.
Refresh Preview	Refreshes the preview data.
Properties	Displays the core query properties.
Advanced Editor	Displays the "M" language editor. This is explained in Chapter 14.
Manage	Lets you delete, duplicate, or reference a query.
Choose Columns	Lets you select the columns to retain from all the columns available in the source data.
Remove Columns	Lets you remove one or more columns.
Keep Rows	Keeps the specified number of rows at the top of the table.
Remove Rows	Removes a specified number of rows from the top of the data table.
Sort	Sorts the table using the selected column as the sort key.
Split Column	Splits a column into one or many columns at a specified delimiter or after a specified number of characters.
Group By	Groups the table using a specified set of columns and aggregates any numeric columns for this grouping.
Data Type	Applies the chosen data type to the column.
Use First Row as Headers	Uses the first row as the column titles.
Replace Values	Carries out a search-and-replace operation on the data in a column or columns. This can affect all or part of the data in a column.

(continued)

Table 7-1. (*continued*)

Option	Description
Merge Queries	Joins a second query table to the current query results and either aggregates or adds data from the second to the first. This is explained in Chapter 9.
Append Queries	Adds the data from another query to the current query in the current Power BI Desktop file. This is explained in Chapter 9.
Combine Files	Adds the data from a series of similarly structured text files into a single table. This is explained in Chapter 12.

The Transform Ribbon

The Transform ribbon, as its name implies, contains a wealth of functions that can help you to transform your data. The various options it contains are explained in Figure 7-6 and Table 7-2.

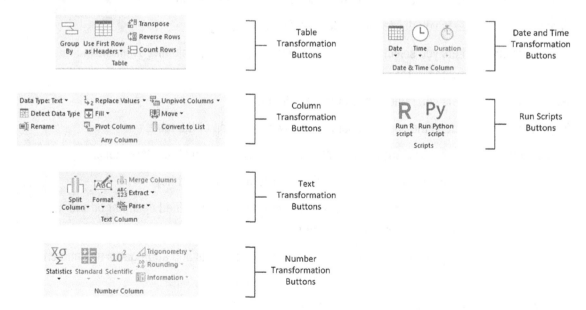

Figure 7-6. *The Query Editor Transform ribbon*

Table 7-2. *Query Editor Transform Ribbon Options*

Option	Description
Group By	Groups the table using a specified set of columns; aggregates any numeric columns for this grouping.
Use First Row as Headers	Uses the first row as the column titles.
Transpose	Transforms the columns into rows and the rows into columns.
Reverse Rows	Displays the source data in reverse order, showing the final rows at the top of the window.
Count Rows	Counts the rows in the table and replaces the data with the row count.
Data Type	Applies the chosen data type to the column.
Detect Data Type	Detects what Power Query thinks is the correct data type to apply to multiple columns.
Rename	Renames a column.
Replace Values	Carries out a search-and-replace operation inside a column, replacing a specified value with another value.
Fill	Copies the data from cells above or below into empty cells in the column.
Pivot Column	Creates a new set of columns using the data in the selected column as the column titles.
Unpivot Columns	Takes the values in a set of columns and unpivots the data, creating new columns using the column headers as the descriptive elements.
Move	Moves a column.
Convert to List	Converts the contents of a column to a list. This can be used, for instance, as query parameters. You will learn this in Chapter 13.
Split Column	Splits a column into one or many columns at a specified delimiter or after a specified number of characters.
Format	Modifies the text format of data in a column (uppercase, lowercase, capitalization) or removes trailing spaces.

(*continued*)

Table 7-2. (*continued*)

Option	Description
Merge Columns	Takes the data from several columns and places it in a single column, adding an optional separator character.
Extract	Replaces the data in a column using a defined subset of the current data. You can specify a number of characters to keep from the start or end of the column, set a range of characters beginning at a specified character, or even list the number of characters in the column.
Parse	Creates an XML or JSON document from the contents of an element in a column.
Statistics	Returns the Sum, Average, Maximum, Minimum, Median, Standard Deviation, Count, or Distinct Value Count for all the values in the column.
Standard	Carries out a basic mathematical calculation (add, subtract, divide, multiply, integer-divide, or return the remainder) using a value that you specify applied to each cell in the column.
Scientific	Carries out a basic scientific calculation (square, cube, power of n, square root, exponent, logarithm, or factorial) for each cell in the column.
Trigonometry	Carries out a basic trigonometric calculation (Sine, Cosine, Tangent, ArcSine, ArcCosine, or ArcTangent) using a value that you specify applied to each cell in the column.
Rounding	Rounds the values in the column either to the next integer (up or down) or to a specified factor.
Information	Replaces the value in the column with simple information: Is Odd, Is Even, or Positive/Negative.
Date	Isolates an element (day, month, year, etc.) from a date value in a column.
Time	Isolates an element (hour, minute, second, etc.) from a date/time or time value in a column.
Duration	Calculates the duration from a value that can be interpreted as a duration in days, hours, minutes, and so forth.
Run R Script	Runs an "R" script to transform data.
Run Python Script	Runs a Python script to transform data.

The Add Column Ribbon

The Add Column ribbon does a lot more than just add columns. It also contains functions to break columns down into multiple columns and to add columns containing dates and calculations based on existing columns. The various options it contains are explained in Figure 7-7 and Table 7-3.

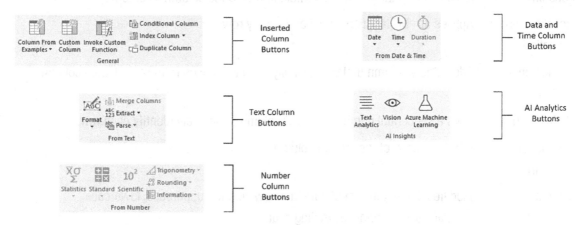

Figure 7-7. *The Query Editor Add Column ribbon*

Table 7-3. *Query Editor Add Column Ribbon Options*

Option	Description
Column From Examples	Lets you use one or more columns as examples to create a new column.
Custom Column	Adds a new column using a formula to create the column's contents.
Invoke Custom Function	Applies an "M" language function to every row.
Conditional Column	Adds a new column that conditionally adds the values from the selected column.
Index Column	Adds a sequential number in a new column to uniquely identify each row.
Duplicate Column	Creates a copy of the current column.
Format	Modifies the text format of data in a new column (uppercase, lowercase, capitalization) or removes trailing spaces.
Merge Columns	Takes the data from several columns and places it in a single column, adding an optional separator character.
Extract	Creates a new column using a defined subset of the current data. You can specify a number of characters to keep from the start or end of the column, set a range of characters beginning at a specified character, or even list the number of characters in the column.
Parse	Creates a new column based on the XML or JSON in a column.
Statistics	Creates a new column that returns the Sum, Average, Maximum, Minimum, Median, Standard Deviation, Count, or Distinct Value Count for all the values in the column.
Standard	Creates a new column that returns a basic mathematical calculation (add, subtract, divide, multiply, integer-divide, or return the remainder) using a value that you specify applied to each cell in the column.
Scientific	Creates a new column that returns a basic scientific calculation (square, cube, power of n, square root, exponent, logarithm, or factorial) for each cell in the column.

(continued)

Table 7-3. (*continued*)

Option	Description
Trigonometry	Creates a new column that returns a basic trigonometric calculation (Sine, Cosine, Tangent, ArcSine, ArcCosine, or ArcTangent) using a value that you specify applied to each cell in the column.
Rounding	Rounds the values in a new column either to the next integer (up or down) or to a specified factor.
Information	Replaces the value in the column with simple information: Is Odd, Is Even, or Positive/Negative.
Date	Isolates an element (day, month, year, etc.) from a date value in a new column.
Time	Isolates an element (hour, minute, second, etc.) from a date/time or time value in a new column.
Duration	Calculates the duration from a value that can be interpreted as a duration in days, hours, minutes, and seconds in a new column.

Note   Several of the options in the Add Column ribbon are also available in the Transform ribbon. These options will carry out the same tasks with one major difference: they transform data in place if you apply the option from the Transform ribbon—whereas they keep the original data and add a new column if you apply the option from the Add Column ribbon.

The View Ribbon

Until now, we have concentrated our attention on Power Query Home, Transform, and Add Column ribbons. This is for the good and simple reason that these ribbons are where nearly all the action takes place. There is, however, a fourth essential Power Query ribbon—the View ribbon. The buttons that it contains are shown in Figure 7-8, and the options are explained in Table 7-4.

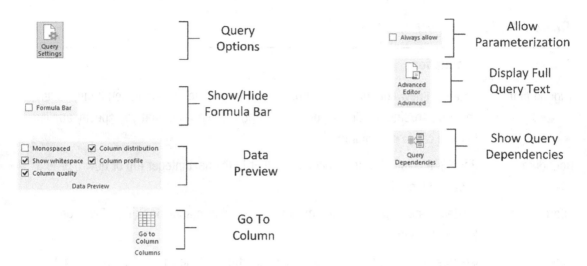

Figure 7-8. *The Power BI Desktop View ribbon*

Table 7-4. *Power BI Desktop View Ribbon Options*

Option	Description
Query Settings	Displays or hides the Query Settings pane at the right of the Power BI Desktop window. This includes the Applied Steps list.
Formula Bar	Shows or hides the formula bar containing the M language code for a transformation step.
Monospaced	Displays previews in a monospaced font.
Show whitespace	Displays whitespace and new line characters.
Column quality	Shows column quality characteristics such as the number of error or empty records.
Column distribution	Shows column distribution characteristics such as the number of distinct elements in the column.
Column profile	Shows column profile characteristics such as value distributions.
Go to Column	Allows you to select a specific column. This is useful for tables with many columns.
Always allow	Allows parameterization in data source and transformation dialogs.
Advanced Editor	Displays the Advanced Editor dialog containing all the code for the steps in the query.
Query Dependencies	Displays the sequence of query links and dependencies.

Possibly the only option that is not immediately self-explanatory is the Advanced Editor button. It displays the code for all the transformations in the query as a single block of "M" language script. You will learn more about this in Chapter 14.

Tip   Personally, I find that the Query Settings pane and the Formula Bar are too vital to be removed from the Power Query window when transforming data. Consequently, I tend to leave them visible. If you need the screen real estate, however, then you can always hide them for a while.

The Tools Ribbon

The Tools ribbon contains diagnostic options to help you to debug and understand Power Query steps.

The Tools buttons are shown in Figure 7-9 and outlined in Table 7-5.

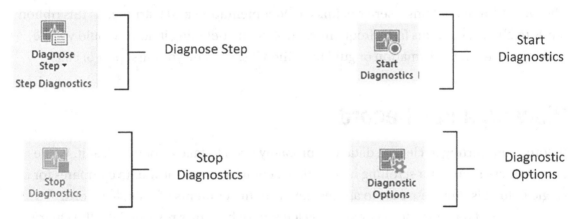

Figure 7-9. *The Tools ribbon*

Table 7-5. *Power BI Desktop Tools Ribbon Options*

Option	Description
Diagnose Step	Allows you to evaluate and diagnose the selected query step
Start Diagnostics	Starts recording traces for query diagnostics
Stop Diagnostics	Stops recording traces for query diagnostics
Diagnostic Options	Displays the diagnostic options screen

The Help Ribbon

The Help ribbon contains a series of links to documentation and learning. As this ribbon tends to change contents fairly frequently, I will not be detailing it here. Should you be looking for further information or guidance, then it should be your first port of call.

Viewing a Full Record

Before even starting to cleanse data, you probably need to take a good look at it. While Power Query is great for scrolling up and down columns to see how data compares for a single field, it is often less easy to appreciate the entire contents of a single record.

So to avoid having to scroll frenetically left and right across rows of data, the Query Editor has another brilliantly simple solution. If you click a row (or more specifically, the number of a row in the grid on the left), Power Query will display the contents of an entire record in a single window under the dataset. You can see an example of this in Figure 7-10, which uses the Power BI Desktop file Example1.pbix.

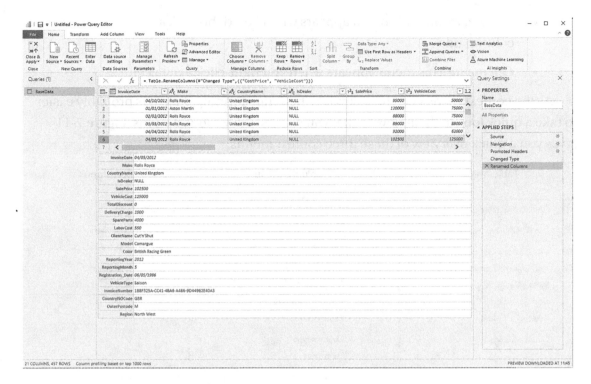

Figure 7-10. *Viewing a full record*

Note   You can alter the relative height of the recordset and dataset windows simply by dragging the gray separator line between the upper and lower windows up or down.

Power Query Context Menus

As is normal for Windows programs, Power Query makes full use of context (or "right-click") menus as an alternative to using the ribbons. When transforming datasets, there are three main context menus that you will probably find yourself using:

- *Table menu*: This menu appears when you click or right-click the top corner of the grid containing the data.

- *Column menu*: This menu appears when you right-click a column title.

- *Cell menu*: This menu appears when you right-click a data cell.

As the context menus can be essential when transforming data, it is probably easier to take a quick look at them now so that you can see the various options. Figure 7-11 gives you a quick overview of these three context menus.

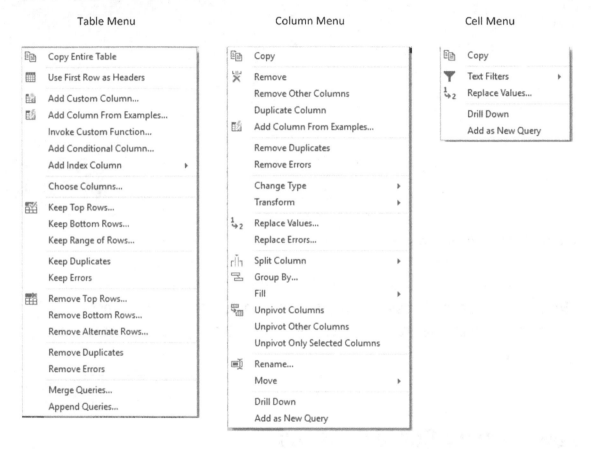

Figure 7-11. *Power Query context menus*

Because the options that are available in the context menus are explained throughout this and the following five chapters, I will not explain them all in detail here.

Note   The cell context menu will reflect the data type of the cell in the filter option. So a numeric cell will have the option "Number filters."

Conclusion

This chapter introduced you to Power Query—the tool that is integrated into Power BI Desktop that allows you to transform and cleanse source data.

You saw that you can switch between the Power BI Desktop dashboard interface and Power Query with one click at any time. Indeed, you learned that you can transform data before loading it—or add data first and then flip over to Power Query to add further transformations to your data.

Now that you know the basic approaches, it is time to examine many of the detailed techniques that you may need to apply to cleanse and restructure raw data. This will be the subject of the next five chapters.

CHAPTER 8

Structuring Data

So far in this book you have discovered how to load data from a multitude of sources. This means that you can now access source data from a vast range of applications and file types. Indeed, you can now build on the knowledge that you have acquired in the first five chapters to load data from just about anything that Power BI can connect to.

The time has come to see how you can take the raw data that you have loaded and cleanse and transform it so that it becomes a valid and reliable source for your dashboards. Consequently, this chapter introduces you to the core data transformation techniques that you can apply to shape each individual dataset that you have loaded. The transformations that I have gathered under this heading include

- Renaming, removing, and reordering columns

- Removing groups or sets of rows

- Deduplicating datasets

- Merging columns

- Sorting the data

- Excluding records by filtering the data

- Grouping records

In this chapter, I will also use a set of example files that you can find on the Apress website. If you have followed the instructions in Appendix A, then these files are in the C:\PowerBiDesktopSamples folder.

© Adam Aspin 2022
A. Aspin, *Pro Data Mashup for Power BI*, https://doi.org/10.1007/978-1-4842-8578-7_8

Dataset Shaping

In this chapter, you will be working in Power Query. So for argument's sake, let's assume that you have opened the `C:\PowerBiDesktopSamples\Example1.pbix` file from the sample data directory and that you have clicked the Transform Data button to display the Power BI Desktop Query Editor. What can you do to the BaseData dataset that is now visible? It is time to take a look at some of the core techniques that you can apply to shape the initial dataset.

I have grouped these techniques together as they affect the initial size and shape of the data. Also, it is generally not only good practice but also easier for you, the data modeler, if you begin by excluding any rows and columns that you do not need. I also find it easier to understand datasets if the columns are logically laid out and given comprehensible names from the start. All in all, this makes working with the data easier in the long run.

In this chapter, you will be using a series of available options from both the Home and Transform ribbons. As the sheer range of available data transformations can seem overwhelming at first, you need to make sure that you are looking in the right ribbon for the menu option that you wish to apply.

Note   Many—if not all—of the transformations that you will learn in this chapter require you to start the transformation step immediately after opening Power Query. So if you are practicing the steps sequentially, it is best to just delete the last step in the "Applied Steps" list to revert things to their previous state. Alternatively, you can close Power BI Desktop without saving any changes and then reopen it before trying out each data transformation.

To allow you to concentrate on discovering, appreciating, and learning the range of data transformation options that are available in Power Query, I will be using only Excel and text/CSV files as sample data in this and the following six chapters. All the Power Query techniques, however, can be applied to *any* source data wherever it comes from.

Renaming Columns

Although we took a quick look at renaming columns in the previous chapter, let's look at this technique again in more detail. I admit that renaming columns is not actually modifying the form of the data table. However, when dealing with data, I consider it vital to have all data clearly identifiable. This implies meaningful column names being applied to each column. After all, you will be using them as the basis for your dashboards. Consequently, I consider this modification to be fundamental to the shape of the data and also as an essential best practice when importing source data.

To rename a column:

1. Click inside the column that you want to rename.

2. Click Transform to activate the Transform ribbon.

3. Click the Rename button. The column name will be highlighted.

4. Enter the new name or edit the existing name.

5. Press Enter or click outside the column title.

The column will now have a new title. The Applied Steps list on the right will now contain another element, Renamed Columns. This step will be highlighted.

Note   As an alternative to using the Transform ribbon, you can right-click the column title and select Rename. You can just double-click the column name, as well, to edit the name. Moreover, you can also change field names in Power BI Desktop—which will apply the changes behind the scenes in Power Query as well.

Reordering Columns

Power BI Desktop will load data as it is defined in the data source. Consequently, the column sequence will be entirely dependent on the source data (or by a SQL query if you used a source database, as described in Chapter 3). This column order need not be definitive, however, and you can reorder the columns if that helps you understand and deal with the data. Do the following to change column order:

1. Click the header of the column you want to move.

2. Drag the column left or right to its new position. You will see the
   column title slide laterally through the column titles as you do
   this, and a thicker gray line will indicate where the column will be
   placed once you release the mouse button. Reordered Columns
   will appear in the Applied Steps list.

Figure 8-1 shows this operation.

InvoiceDate	Make	CountryName	Make	SalePrice
04/10/2012	Rolls Royce	United Kingdom	NULL	95000
01/01/2012	Aston Martin	United Kingdom	NULL	120000
02/02/2012	Rolls Royce	United Kingdom	NULL	88000
03/03/2012	Rolls Royce	United Kingdom	NULL	89000
04/04/2012	Rolls Royce	United Kingdom	NULL	92000
04/05/2012	Rolls Royce	United Kingdom	NULL	102500
04/06/2012	Aston Martin	United Kingdom	NULL	110000
04/07/2012	Aston Martin	France	NULL	125000
04/08/2012	Aston Martin	United Kingdom	NULL	130000
04/09/2012	Aston Martin	United Kingdom	NULL	75000
04/09/2012	Aston Martin	United Kingdom	NULL	68500
04/11/2012	Aston Martin	France	NULL	95000
04/11/2012	Aston Martin	United Kingdom	NULL	155000
04/12/2012	Aston Martin	United Kingdom	NULL	95000

Figure 8-1. *Reordering columns*

If your query contains dozens—or even hundreds—of columns, you may find that
dragging a column around can be slow and laborious. Equally, if columns are extremely
wide, it can be difficult to "nudge" a column left or right. Power BI Desktop can come to
your aid in these circumstances with the Move button in the Transform ribbon. Clicking
this button gives you the menu options that are outlined in Table 8-1.

Table 8-1. *Move Button Options*

Option	Description
Left	Moves the currently selected column to the left of the column on its immediate left
Right	Moves the currently selected column to the right of the column on its immediate right
To Beginning	Moves the currently selected column to the left of all the columns in the query
To End	Moves the currently selected column to the right of all the columns in the query

The Move command also works on a set of columns that you have selected by Ctrl-clicking and/or Shift-clicking. Indeed, you can move a selection of columns that is not contiguous if you need to. Indeed, if they're not already contiguous, a "move" operation on multiple columns *will* make them contiguous. In other words, the columns don't move the same number of steps relative to their original position–they're essentially grouped together and then inserted into the new position.

Note   You need to select a column (or a set of columns) before clicking the Move button. If you do not, then the first time that you use Move, Power BI Desktop Query selects the column(s) but does not move them.

Removing Columns

So how do you remove a column or series of columns from the dataset? Like this:

1. Click inside the column you want to delete, or if you want to delete several columns at once, Ctrl-click the titles of the columns that you want to delete.

2. Click the Remove Columns button in the Home ribbon. The column(s) will be deleted, and Removed Columns will be the latest element in the Applied Steps list.

Note   Removing a column has no effect on the source data. As is the case with all data transformation in Power Query, only the data in Power BI Desktop is affected by the changes that you make.

When working with imported datasets over which you have had no control, you may frequently find that you only need a few columns of a large data table. If this is the case, you will soon get tired of Ctrl-clicking numerous columns to select those you want to remove. Power BI Desktop has an alternative method. Just select the columns you want to keep and delete the others. To do this:

1. Ctrl-click the titles of the columns that you want to keep.

2. Click the small triangle in the Remove Columns button in the Home ribbon. Select Remove Other Columns from the menu. All unselected columns will be deleted, and Removed Other Columns will be added to the Applied Steps list.

When selecting a contiguous range of columns to remove or keep, you can use the standard Windows Shift-click technique to select from the first to the last column in the block of columns that you want to select.

Note   Both of these options for removing columns are also available from the context menu, if you prefer. It shows Remove (or Remove Columns, if there are several columns selected) when deleting columns, as well as Remove Other Columns if you right-click a column title.

Choosing Columns

If you prefer not to scroll through a wide dataset yet still need to select a subset of columns as the basis for your reports, then there is another way to define the collection of fields that you want to use. You can choose the columns that you want to keep (and, by definition, those that you want to exclude) like this:

1. In the Home ribbon of Power Query, click the Choose Columns button.

2. Click (Select All Columns) to deselect the entire collection of columns in the dataset.

3. Select the columns Make, Color, and SalePrice. The Choose Columns dialog will look like the one in Figure 8-2.

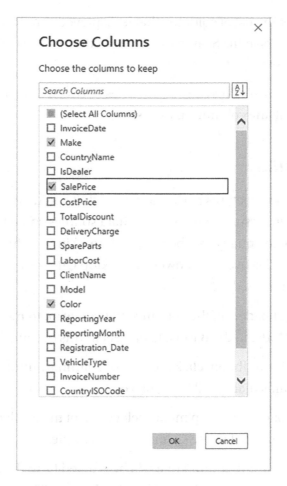

Figure 8-2. *The Choose Columns dialog*

4. Click OK. Power Query will only display the columns that you
    selected.

The Choose Columns dialog comes with a couple of extra functions that you might
find useful when choosing the set of columns that you want to work with:

- You can sort the column list in alphabetical order (or, indeed,
  revert to the original order) by clicking the Sort icon (the small A-Z)
  at the top right of the Choose Columns dialog and selecting the
  required option.

- You can filter the list of columns that is displayed simply by entering a few characters in the Search Columns field at the top of the dialog and then pressing Enter.

- The (Select All Columns) option switches between selecting and deselecting all the columns in the list.

Merging Columns

Source data is not always exactly as perfect as you wish it could be (and that is sometimes a massive understatement). Certain data sources could have data spread over many columns that could equally well be merged into a single column. So it probably comes as no surprise to discover that Power BI Desktop Query can carry out this kind of operation too. Here is how to do it:

1. Ctrl-click the headers of the columns that you want to merge (Make and Model in the BaseData dataset in this example).

2. In the Transform ribbon, click the Merge Columns button. The Merge Columns dialog will be displayed.

3. From the Separator pop-up menu, select one of the available separator elements. I chose Colon in this example.

4. Enter a name for the column that will be created from the two original columns (I am calling it MakeAndModel). The dialog should look like Figure 8-3.

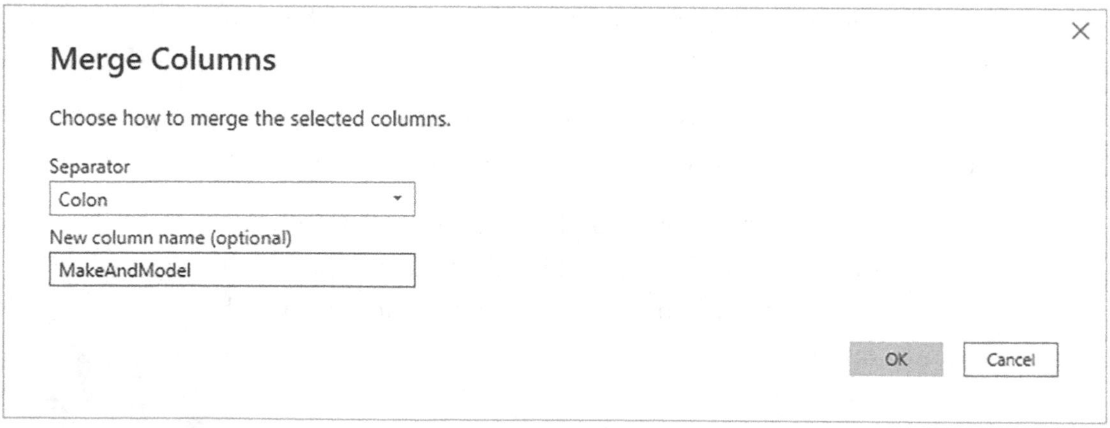

Figure 8-3. The Merge Columns dialog

5. Click OK. The columns that you selected will be replaced by a single column containing the data from all the selected columns, as shown in Figure 8-4.

A^B_C ClientName	▼	A^B_C MakeAndModel	▼	A^B_C Color	▼
Aldo Motors		Rolls Royce:Camargue		Red	
Honest John		Aston Martin:DBS		Blue	
Bright Orange		Rolls Royce:Silver Ghost		Green	
Honest John		Rolls Royce:Silver Ghost		Blue	
Wheels'R'Us		Rolls Royce:Camargue		Canary Yellow	
Cut'n'Shut		Rolls Royce:Camargue		British Racing Green	
Bright Orange		Aston Martin:DBS		Dark Purple	
Les Arnaqueurs		Aston Martin:DB7		Red	
Aldo Motors		Aston Martin:DB9		Blue	
Aldo Motors		Aston Martin:DB9		Silver	
Aldo Motors		Aston Martin:DB4		Night Blue	
Les Arnaqueurs		Aston Martin:Vantage		Canary Yellow	
Wheels'R'Us		Aston Martin:Vanquish		Night Blue	
Bright Orange		Aston Martin:Rapide		Black	
Cut'n'Shut		Aston Martin:Zagato		British Racing Green	
Les Arnaqueurs		Rolls Royce:Silver Ghost		Canary Yellow	
Wheels'R'Us		Rolls Royce:Wraith		Silver	

Figure 8-4. *The result of merging columns*

Note   The replacement column will be named "Merged" if you did not provide a new name in step 4. In any case, you can always rename the column to something more appropriate after merging columns.

I need to make a few comments about this process:

- You can select as many columns as you want when merging columns.

- If you do not give the resulting column a name in the Merge Columns dialog, it will simply be renamed Merged. You can always rename it later if you want.

- The order in which you select the columns affects the way that the data is merged. So always begin by selecting the column whose data must appear at the left of the merged column, then the column whose data should be next, and so forth. You do not have to select columns in the order that they initially appeared in the dataset.

- If you do not want to use any of the standard separators that Power BI Desktop Query suggests, you can always define your own. Just select Custom in the pop-up menu in the Merge Columns dialog. A new box will appear in the dialog, in which you can enter your choice of separator. This can be composed of several characters if you really want.

- Merging columns from the Transform ribbon removes all the selected columns and replaces them with a single column. The same option is also available from the Add Column ribbon—only in this case, this operation *adds* a new column and leaves the original columns in the dataset.

Note   This option is also available from the context menu if you right-click one of the selected column titles.

The available merge separators are described in Table 8-2.

Table 8-2. *Merge Separators*

Option	Description
Colon	Uses the colon (:) as the separator
Comma	Uses the comma (,) as the separator
Equals Sign	Uses the equals sign (=) as the separator
Semi-Colon	Uses the semicolon (;) as the separator
Space	Uses the space ( ) as the separator
Tab	Uses the tab character as the separator
Custom	Lets you enter a custom separator

Tip   You can split, remove, and duplicate columns using the context menu if you prefer. Just remember to right-click any of the selected column titles to display the correct context menu.

Going to a Specific Column

Power BI Desktop can load datasets that contain hundreds of columns. As scrolling left and right across dozens of columns can be more than a little frustrating, you can always jump to a specific column at any time.

1. In the Home ribbon of Power Query, click the small triangle at the bottom of the Choose Columns button. Select Go to Column. The Go to Column dialog will appear, as shown in Figure 8-5.

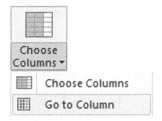

Figure 8-5.   *The Go to Column menu options*

2. Select the column you want to move to. The dialog will look like Figure 8-6.

Figure 8-6. *The Go to Column dialog*

    3.   Click OK. Power BI Desktop will select the chosen column.

Tip   If you prefer, you can double-click a column name in the Go to Column dialog to move to the chosen column. What is more, if you switch to the View menu, there is a Go to column button there, as well.

Removing Records

You may not always need *all* the data that you have loaded into a Power BI Desktop query. There could be several possible reasons for this:

- You are taking a first look at the data and you only need a sample to get an idea of what the data is like.

- The data contains records that you clearly do not need and that you can easily identify from the start.

- You are testing data cleansing and you want a smaller dataset to really speed up the development of a complex data extraction and transformation process.

- You want to analyze a reduced dataset to extrapolate theses and inferences, and to save analysis on a full dataset for later, or even use a more industrial-strength toolset such as SQL Server Integration Services or Azure Data Factory.

To allow you to reduce the size of the dataset, Power BI Desktop proposes two basic approaches out of the box:

- Keep certain rows

- Remove certain rows

Inevitably, the technique that you adopt will depend on the circumstances. If it is easier to specify the rows to sample by inclusion, then the keep certain rows approach is the best option to take. Inversely, if you want to proceed by exclusion, then the remove certain rows technique is best. Let's look at each of these in turn.

Note   Removing of rows should be used with caution if the data source doesn't provide the rows in a deterministic order between refreshes. Some source data connections may not guarantee row order between data refreshes, and in cases like these, "removing top/bottom N rows" may be removing different rows each time.

Rows

This approach lets you specify the rows that you want to continue using. It is based on the application of one of the following three choices:

- Keep the top *n* records.

- Keep the bottom *n* records.

- Keep a specified range of records—that is, keep *n* records every *y* records.

Most of these techniques are very similar, so let's start by imagining that you want to keep the top 50 records in the sample `C:\PowerBiDesktopSamples\Example1.pbix` file.

1. In the Power Query Home ribbon, click the Keep Rows button. The menu will appear.

2. Select Keep Top Rows. The Keep Top Rows dialog will appear.

3. Enter **50** in the "Number of rows" box, as shown in Figure 8-7.

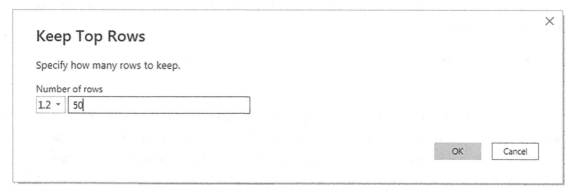

Figure 8-7. *The Keep Top Rows dialog*

4. Click OK. All but the first 50 records are deleted, and Kept First Rows is added to the Applied Steps list.

To keep the bottom *n* rows, the technique is virtually identical. Follow the steps in the previous example, but select *Keep Bottom Rows* in step 2. In this case, the Applied Steps list displays Kept Last Rows.

To keep a range of records, you need to specify a starting record and the number of records to keep from then on. For instance, suppose that you wish to lose the first 10 records but keep the following 25. This is how to go about it:

1. In the Home ribbon, click the Keep Rows button. The Keep Rows menu options will appear as shown in Figure 8-8.

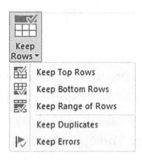

Figure 8-8. *The Keep Rows menu options*

2. Select Keep Range of Rows. The Keep Range of Rows dialog will appear.

3. Enter **11** in the "First row" box.

4. Enter **25** in the "Number of rows" box, as shown in Figure 8-9.

Keep Range of Rows

Specify the range of rows to keep.

First row
| 1.2 ▾ | 11 |

Number of rows
| 1.2 ▾ | 25 |

OK   Cancel

Figure 8-9. *The Keep Range of Rows dialog*

5. Click OK. All but records 1–10 and 36 to the end are deleted, and Kept Range of Rows is added to the Applied Steps list.

Note   You may have noticed that this dialog—like many others—contains a pop-up menu to the left of the fields where you enter values. This pop-up menu allows you to *parameterize* the value that you are currently entering manually. Parameters are explained in Chapter 13.

Removing Rows

Removing rows is a nearly identical process to the one you just used to keep rows. As removing the top or bottom *n* rows is highly similar, I will not go through it in detail. All you have to do is click the Remove Rows button in the Home ribbon and follow the process as if you were keeping rows. The Applied Steps list will read Removed Top Rows or Removed Bottom Rows in this case, and rows will be removed instead of being kept in the dataset, of course.

The remove rows approach does have one very useful option that can be applied as a sampling technique. It allows you to remove one or more records every few records to produce a subset of the source data. To do this, you need to do the following:

1. Click the Remove Rows button in the Query window Home ribbon. The menu will appear as shown in Figure 8-10.

Figure 8-10.   *The Remove Rows menu options*

2. Select Remove Alternate Rows. The Remove Alternate Rows dialog will appear.

3. Enter **10** as the First row to remove.

4. Enter **2** as the Number of rows to remove.

5. Enter **10** as the Number of rows to keep.

The dialog will look like Figure 8-11.

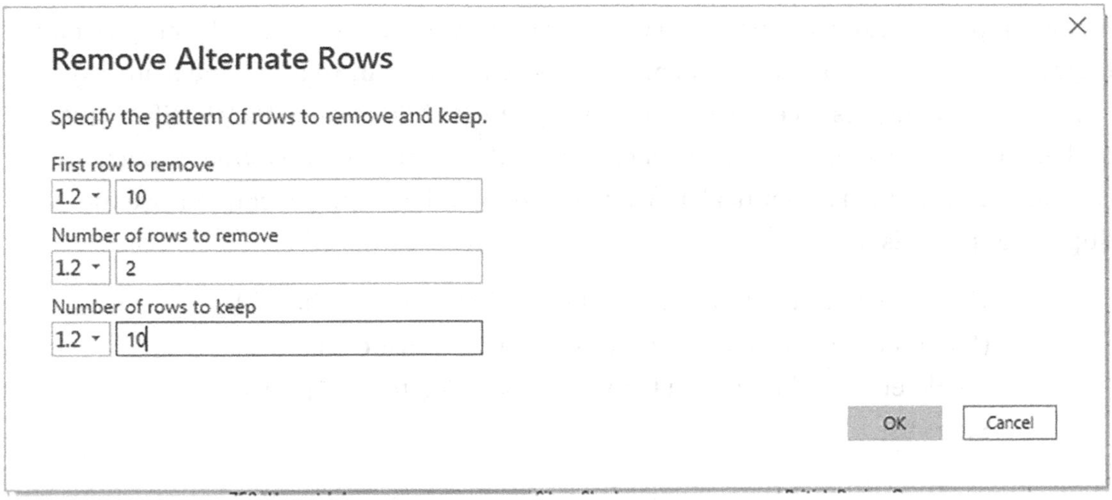

Remove Alternate Rows

Specify the pattern of rows to remove and keep.

First row to remove
1.2 ▾   10

Number of rows to remove
1.2 ▾   2

Number of rows to keep
1.2 ▾   10

OK   Cancel

Figure 8-11. *The Remove Alternate Rows dialog*

6. Click OK. All but the records matching the pattern you entered in
    the dialog are removed. Removed Alternate Rows is then added to
    the Applied Steps list.

Note   If you are really determined to extract a sample that you consider to
be representative of the key data, then you can always filter the data before
subsetting it to exclude any outliers. Filtering data is explained later in this chapter.

Removing Blank Rows

If your source data contains completely blank (empty) rows, you can delete these as
follows:

1. Click the Remove Rows button in the Query window Home
    ribbon. The menu will appear.

2. Select Remove Blank Rows.

This results in empty rows being deleted. Removed Blank Rows is then added to the
Applied Steps list.

Removing Duplicate Records

As mentioned previously—and as you will inevitably discover—an external source of data might not be quite as perfect as you might hope. One of the most annoying features of poor data is the presence of duplicates. These are insidious since they falsify results and are not always visible. If you suspect that the data table contains *strict* duplicates (i.e., where every field is identical in two or more records), then you can remove the duplicates like this:

1. Click the Remove Duplicates in the pop-up menu for the table (this is at the top left of the table grid). All duplicate records are deleted, and Removed Duplicates is added to the Applied Steps list.

Note   I must stress that this approach will only remove *completely* identical records where every element of every column is strictly identical in the duplicate rows. If two records have just one different character or a number but everything else is identical, then they are *not* considered duplicates by Power Query. Alternatively, if you want to isolate and examine the duplicate records, then you can display only completely identical records by selecting Keep Duplicates from the pop-up menu for the table.

So if you suspect or are sure that the data table you are dealing with contains duplicates, what are the practical solutions? This can be a real conundrum, but there are some basic techniques that you can apply:

- Remove all columns that you are sure you will not be using later in the data-handling process. This way, Power BI Desktop will only be asked to compare essential data across potentially duplicate records.

- Group the data on the core columns (this is explained later in this chapter) and perform a count to determine whether more than one record exists for that combination of values.

> **Note**   As you have seen, Power BI Desktop Query can help you to home in on
> the essential elements in a dataset in just a few clicks. If anything, you need to be
> careful that you are *not removing* valuable data—and consequently skewing your
> analysis—when excluding data from the query.

Sorting Data

Although not strictly a data modification step, sorting an imported table will probably be
something that you want to do at some stage, if only to get a clearer idea of the data that
you are dealing with. Do the following to sort the data:

1. Close the file Example1.pbix (if, indeed, you have been using it
    to test the techniques explained so far in this chapter) without
    saving it.

2. Open the file `C:\PowerBiDesktopSamples\Example1.pbix`. Note
    that you are opening an existing Power BI Desktop file and *not*
    loading data.

3. Click Transform Data to open Power Query.

4. Click inside the column you wish to sort by.

5. Click Sort Ascending (the A/Z icon) or Sort Descending (the Z/A
    icon) in the Home ribbon.

The data is sorted in either alphabetical (smallest to largest) or reverse alphabetical
(largest to smallest) order. If you want to carry out a complex sort operation (i.e., first
by one column and then by another if the first column contains the same element over
several rows), you do this simply by sorting the columns one after another. Power BI
Desktop Query Editor adds a tiny 1, 2, 3, and so on to the right of the column title to
indicate the sort sequence. You can see this in Figure 8-12, where I sorted *first* on the
column Make and *finally* on the column CountryName.

Figure 8-12. *Sorting multiple columns*

Once a column is sorted, the pop-up menu icon will change to a small triangle and an arrow. An upward arrow indicates that the column has been sorted using an ascending (A-Z or smallest to largest) sort, whereas a downward arrow indicates that the column has been sorted using a descending (Z-A or largest to smallest) sort.

Note   It is important to realize that sorting data is merely a technique to help you understand the source data. This will have no effect on how data appears in visuals in Power BI Desktop.

As sorting data is considered part of the data modification process, it also appears in the Applied Steps list as Sorted Rows. If you look closely at the column headings, you will see small numbers "1" and "2" that indicate the sort priority as well as the arrows that indicate that the columns are sorted in ascending order.

Tip   An alternative technique for sorting data is to click the pop-up menu for a column (the downward-facing triangle at the right of a column title) and select Sort Ascending or Sort Descending from the pop-up menu.

Reversing the Row Order

If you find that the data that you are looking at seems upside down (i.e., with the bottom rows at the top and vice versa), you can reverse the row order in a single click, if you want. To do this, do the following:

1.   In the Transform ribbon, click the Reverse Rows button.

The entire dataset will be reversed, and the bottom row will now be the top row.

Undoing a Sort Operation

If you subsequently decide that you do not want to keep your data sorted, you can undo the sort operation at any time, as follows:

1.   Click the sort icon at the right of the name of the column that you used as the basis for the sort operation. The context menu will appear, as you can see in Figure 8-13.

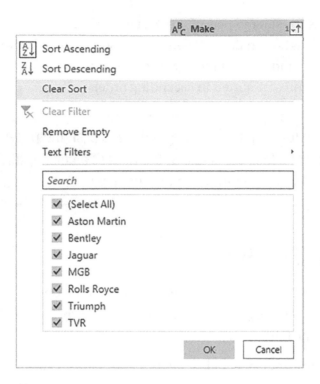

Figure 8-13. *Removing a sort operation*

   2.   Click Clear Sort.

   The sort order that you applied will be removed, and the data will revert to its original row order.

Note   If you sorted the dataset on several columns, removing the sort on the first column will leave the data sorted on the other columns. If all you want to do is undo the sort on the final column in a set of columns used to sort the recordset, then you can clear only the sort operation on this column.

Filtering Data

The most frequently used way of limiting a dataset is, in my experience, the use of filters on the table that you have loaded. Now, I realize that you may be coming to Power BI Desktop after years with Excel, or after some time using Power Pivot, and that the filtering techniques that you are about to see probably look much like the ones you have used in those two tools. However, because it is fundamental to include and exclude appropriate records when loading source data, I prefer to explain Power Query filters in detail, even if this means that certain readers will experience a strong sense of déjà vu.

Here are two basic approaches for filtering data in Power BI Desktop:

- Select one or more specific values from the unique list of elements in the chosen column.

- Define a range of data to include or exclude.

The first option is common to all data types, whether they are text, number, or date/time. The second approach varies according to the data type of the column that you are using to filter data.

Selecting Specific Values

Selecting one or more values present in a column of data is as easy as this (assuming that you are still using the Power BI Desktop file Example1.pbix and are in Power Query):

1. Click a column's pop-up menu. (I used Make in the sample dataset in this example.) The filter menu appears.

2. Check all elements that you want to retain and uncheck all elements that you wish to exclude. In this example, I kept Bentley and Rolls Royce, as shown in Figure 8-14.

Figure 8-14. *A filter menu*

3.   Click OK. The Applied Steps list adds Filtered Rows.

Note   You can deselect all items by clicking the (Select All) check box; reselect all the items by selecting this box again. It follows that if you want to keep only a few elements, it may be faster to unselect all of them first and then only select the ones that you want to keep. If you want to exclude any records without a value in the column that you are filtering on, then select Remove Empty from the filter menu.

Finding Elements in the Filter List

Scrolling up and down in a filter list can get extremely laborious. A fast way of limiting the list to a subset of available elements is to do the following:

1.   Click the pop-up menu for a column. (I use Make in the sample dataset in this example.) The filter menu appears.

2. Enter a letter or a few letters in the Search box. The list shortens with every letter or number that you enter. If you enter **ar**, then the filter popup will look like Figure 8-15.

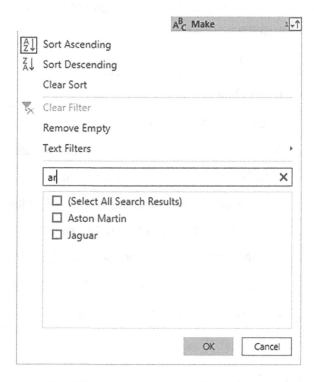

Figure 8-15. *Searching the filter menu*

To remove a filter, all that you have to do is click the cross that appears at the right of the Search box.

Tip   Sometimes, when filtering values, you may see "list may be incomplete" at the bottom of the pop-up menu. In these cases, you may need to click "Load more" to see the values that are missing. This is because Power Query only loads a sample of the source data, and this sample may be too small to be truly representative of the source data.

Filtering Text Ranges

If a column contains text, then you can apply specific options to filter the data. These elements are found in the filter popup of any text-based column in the Text Filters submenu. The choices are given in Table 8-3.

Table 8-3. *Text Filter Options*

Filter Option	Description
Equals	Sets the text that must match the cell contents
Does Not Equal	Sets the text that must *not* match the cell contents
Begins With	Sets the text at the left of the cell contents
Does Not Begin With	Sets the text that must *not* appear at the left of the cell contents
Ends With	Sets the text at the right of the cell contents
Does Not End With	Sets the text that must *not* appear at the right of the cell contents
Contains	Lets you enter a text that will be part of the cell contents
Does Not Contain	Lets you enter a text that will *not* be part of the cell contents

Filtering Numeric Ranges

If a column contains numbers, then there are also specific options that you can apply to filter the data. You'll find these elements in the filter popup of any text-based column in the Number Filters submenu. The choices are given in Table 8-4.

Table 8-4. *Numeric Filter Options*

Filter Option	Description
Equals	Sets the number that must match the cell contents.
Does Not Equal	Sets the number that must *not* match the cell contents.
Greater Than	Cell contents must be greater than this number.
Greater Than Or Equal To	Cell contents must be greater than or equal to this number.
Less Than	Cell contents must be less than this number.
Less Than Or Equal To	Cell contents must be less than or equal to this number.
Between	Cell contents must be between the two numbers that you specify. This enables you to define a *range* of numbers.

Filtering Date and Time Ranges

If a column contains dates or times (or both), then specific options can also be applied to filter the data. These elements are found in the filter popup of any text-based column in the Date/Time Filters submenu. The choices are given in Table 8-5.

Table 8-5. *Date and Time Filter Options*

Filter Element	Description
Equals	Filters data to include only records for the selected date
Before	Filters data to include only records up to the selected date
After	Filters data to include only records after the selected date
Between	Lets you set an upper and a lower date limit to exclude records outside that range
In the Next	Lets you specify a number of days, weeks, months, quarters, or years (or even seconds, minutes, hours) to come
In the Previous	Lets you specify a number of days, weeks, months, quarters, or years (or even seconds, minutes, hours) up to the date

(continued)

Table 8-5. (*continued*)

Filter Element	Description
Is Earliest	Filters data to include only records for the earliest date
Is Latest	Filters data to include only records for the latest date
Is Not Earliest	Filters data to include only records for dates not including the earliest date
Is Not Latest	Filters data to include only records for dates not including the latest date
Day ➤ Tomorrow	Filters data to include only records for the day after the current system date
Day ➤ Today	Filters data to include only records for the current system date
Day ➤ Yesterday	Filters data to include only records for the day before the current system date
Week ➤ Next Week	Filters data to include only records for the next calendar week
Week ➤ This Week	Filters data to include only records for the current calendar week
Week ➤ Last Week	Filters data to include only records for the previous calendar week
Month ➤ Next Month	Filters data to include only records for the next calendar month
Month ➤ This Month	Filters data to include only records for the current calendar month
Month ➤ Last Month	Filters data to include only records for the previous calendar month
Month ➤ Month Name	Filters data to include only records for the specified calendar month
Quarter ➤ Next Quarter	Filters data to include only records for the next quarter
Quarter ➤ This Quarter	Filters data to include only records for the current quarter
Quarter ➤ Last Quarter	Filters data to include only records for the previous quarter
Quarter ➤ Quarter Name	Filters data to include only records for the specified quarter
Year ➤ Next Year	Filters data to include only records for the next year
Year ➤ This Year	Filters data to include only records for the current year
Year ➤ Last Year	Filters data to include only records for the previous year
Year ➤ Year To Date	Filters data to include only records for the calendar year to date

(*continued*)

Table 8-5. (*continued*)

Filter Element	Description
Hour ➤ Next Hour	Filters data to include only records for the next hour
Hour ➤ This Hour	Filters data to include only records for the current hour
Hour ➤ Last Hour	Filters data to include only records for the last hour
Minute ➤ Next Minute	Filters data to include only records for the next minute
Minute ➤ This Minute	Filters data to include only records for the current minute
Minute ➤ Last Minute	Filters data to include only records for the last minute
Second ➤ Next Second	Filters data to include only records for the next second
Second ➤ This Second	Filters data to include only records for the current second
Second ➤ Last Second	Filters data to include only records for the last second
Custom Filter	Lets you set up a specific filter for a chosen date range

Filtering Data

Filtering data uses a very similar approach, whatever the type of filter that is applied. As a simple example, here is how to apply a number filter to the sale price to find vehicles that sold for less than £5,000.00:

1. Click the pop-up menu for the SalePrice column.

2. Click Number Filters. The submenu will appear.

3. Select Less Than. The Filter Rows dialog will be displayed.

4. Enter **5000** in the box next to the "is less than" box, as shown in Figure 8-16.

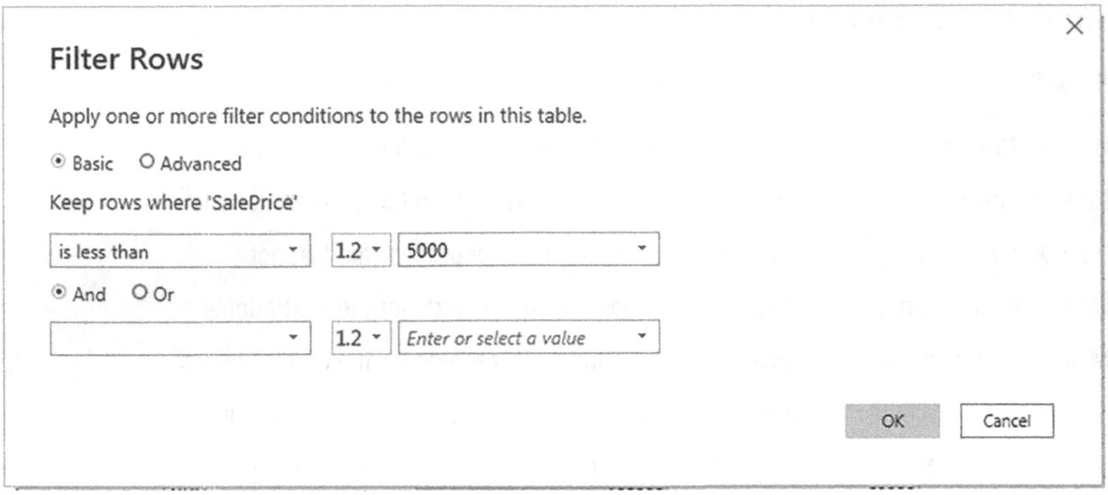

Figure 8-16. *The Filter Rows dialog*

5. Click OK. The dataset only displays rows that conform to the filter that you have defined.

Although extremely simple to apply, filters do require a few comments:

- You can combine up to two elements in a basic filter. These can be mutually inclusive (an AND filter) or they can be an *alternative* (an OR filter).

- You can combine several elements in an advanced filter—as you will learn in the next section.

- You should *not* apply any formatting when entering numbers.

- Any text that you filter on is case-sensitive.

- If you choose the wrong filter, you do not have to cancel and start over. Simply select the correct filter type from the popup in the left-hand boxes in the Filter Rows dialog.

Tip   If you set a filter value that excludes all the records in the table, Power BI Desktop displays an empty table except for the words "This table is empty." You can always remove the filter by clicking the cross to the left of Filtered Rows in the Applied Steps list. This will remove the step and revert the data to its previous state.

Applying Advanced Filters

Should you ever need to be extremely specific when filtering data, you can always use Power BI Desktop's advanced filters. These let you extend the filter elements so that you can include or exclude records to a high level of detail. Here is the procedure:

1. Click the pop-up menu for the SalePrice column.

2. Click Number Filters. The submenu will appear.

3. Select Equals. The Filter Rows dialog will be displayed.

4. Click Advanced.

5. Enter **5000** as the value for the first filter element in the dialog.

6. Select Or from the popup as the filter type for the second filter element.

7. Select Equals as the operator.

8. Enter **89000** as the value for the second filter in the dialog.

9. Click Add clause. A new filter element will be added to the dialog under the existing elements.

10. Select Or from the popup as the filter type for the third filter element.

11. Select Equals as the operator.

12. Enter **178500** as the value for the third filter element in the dialog. The Filter dialog will look like the one shown in Figure 8-17.

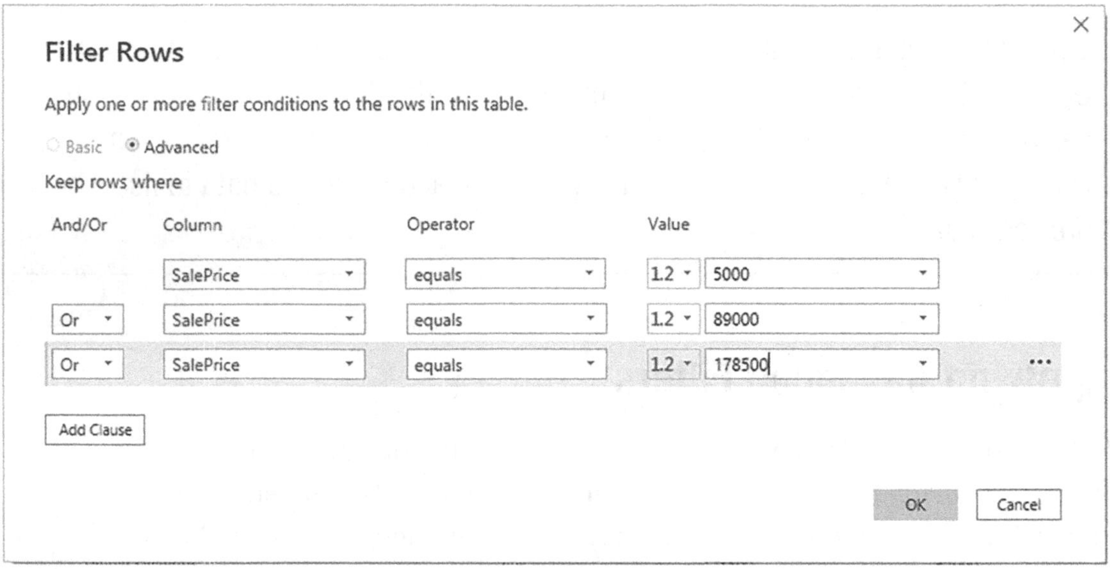

Figure 8-17. *Advanced filters*

13. Click OK. Only records containing the figures that you entered in the Filter dialog will be displayed in the Power BI Desktop Query Editor.

I would like to finish on the subject of filters with a few comments:

- In the Advanced Filter dialog, you can "mix and match" columns and operators to achieve the filter result that you are looking for.

- You can also order the sequence of filters if you ever need to. To do this, simply click inside a filter row and it will appear with a gray background (like the third filter in Figure 8-17). Then click the ellipses at the right of the filter row and select Move Up or Move Down from the pop-up menu. You can see this in Figure 8-18.

Figure 8-18. *Ordering filters*

- To delete a filter, click the ellipses at the right of the filter row and select Delete.

Note   To repeat a truly fundamental point, when you are dealing with really large datasets, you may find that a filter does not always show all the available values from the source data. This is because Power Query has loaded only a *sample* subset of the data. In cases like these, you will see an alert in the filter pop-up menu and a "Load more" link. Clicking this link will force Power Query to reload a larger sample set of data. However, memory restrictions may prevent it from loading all the data that you need. In cases like this, you should consider modifying the source query if possible so that it brings back a representative dataset that can fit into memory.

Excluding Rows Where a Value Is Missing

In some cases, you may wish to exclude records where a specific column does not contain a value. Suppose that you do not want to include data where there is no sale price.

1. Click the pop-up menu for the SalePrice column.

2. Select Remove Empty. You can see this in Figure 8-19.

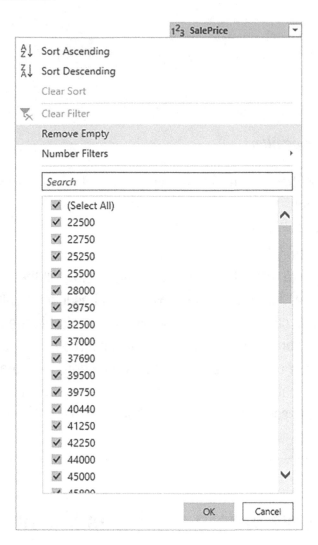

Figure 8-19. Filtering out empty records

Grouping Records

At times, you will need to transform your original data in an extreme way—by grouping the data. This is very different from filtering data, removing duplicates, or cleansing the contents of columns. When you group data, you are altering the structure of the dataset to "roll up" records where you do the following:

- Define the attribute columns that will become the unique elements in the grouped data table.

- Specify which aggregations are applied to any numeric columns included in the grouped table.

Grouping is frequently an extremely selective operation. This is inevitable, since the fewer attribute (i.e., nonnumeric) columns you choose to group on, the fewer records you are likely to include in the grouped table. However, this will always depend on the particular dataset you are dealing with, and grouping data efficiently is always a matter of flair, practice, and good, old-fashioned trial and error.

Simple Groups

To understand how grouping works—and how it can radically alter the structure of your dataset—let's see a simple example of row grouping in action:

1. In the sample file Example1.pbix (in Power Query), click inside the Make column.

2. In either the Home ribbon or Transform ribbon, click the Group By button. The Group By dialog will appear, looking like the one in Figure 8-20.

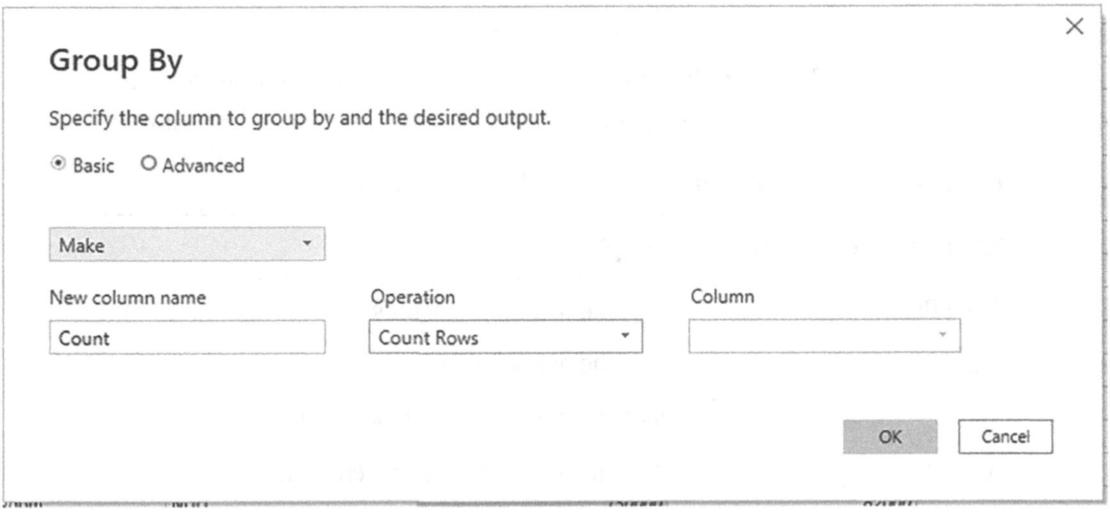

Figure 8-20. *Simple grouping*

3. Click OK. The dataset will now only contain the list of makes of vehicle and the number of records for each make. You can see this in Figure 8-21.

	A^B_C Make	1²₃ Count
1	Rolls Royce	63
2	Aston Martin	110
3	Jaguar	129
4	Bentley	71
5	TVR	14
6	MGB	36
7	Triumph	34

Figure 8-21. Simple grouping output

Power BI Desktop will add a step named Grouped Rows to the Applied Steps list when you apply grouping to a dataset.

Note   The best way to cancel a grouping operation is to delete the Grouped Rows step in the Applied Steps list.

Although Power BI Desktop defaults to counting rows, there are several other operations that you can apply when grouping data. These are outlined in Table 8-6.

Table 8-6. Aggregation Operations When Grouping

Aggregation Operation	Description
Count Rows	Counts the number of records
Count Distinct Rows	Counts the number of unique records
Sum	Returns the total for a numeric column
Average	Returns the average for a numeric column
Median	Returns the median value of a numeric column
Min	Returns the minimum value of a numeric column
Max	Returns the maximum value of a numeric column
All Rows	Creates a table of records for each grouped element

Note   For aggregations other than Count Rows and All Rows, you can select the column to aggregate from the Column popup in the Group By dialog. This only applies to numeric columns.

Complex Groups

Power BI Desktop can help you shape your datasets in more advanced ways by creating more complex data groupings. As an example, you could try out the following to group by make and model and add columns showing the total sales value and the average cost:

1. (Re)open the sample file `C:\PowerBiDesktopSamples\`
    `Example1.pbix`.

2. Click Transform Data to open Power Query.

3. Select the following columns (by Ctrl-clicking the column headers):

    a. Make

    b. CountryName

4. In either the Home ribbon or Transform ribbon, click Group By.

5. In the New Column Name box, enter **TotalSales**.

6. Select Sum as the operation.

7. Choose SalePrice as the source column in the Column pop-up list.

8. Click the Add Aggregation button and repeat the operation; only this time, use the following:

    a. *New Column Name*: **AverageCost**

    b. *Operation*: Average

    c. *Column*: CostPrice

The Group By dialog should look like the one in Figure 8-22.

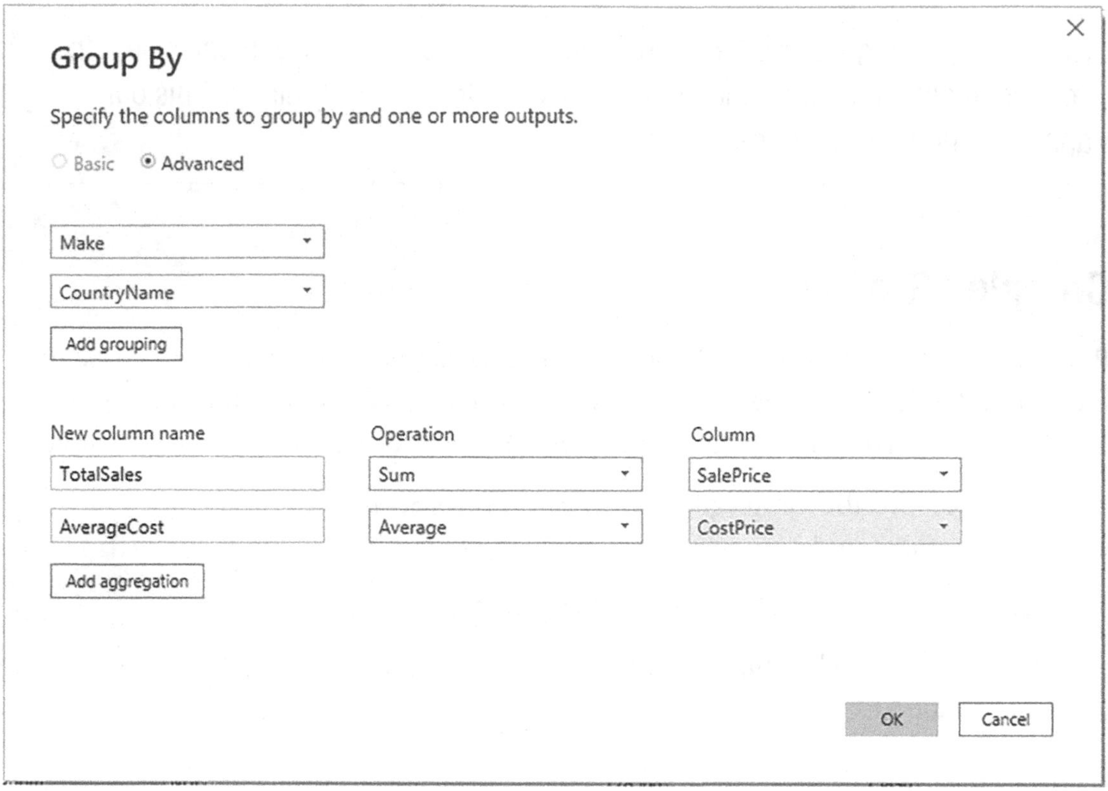

Figure 8-22. *The Group By dialog*

9.   Click OK. All columns, other than those that you specified in
the Group By dialog, are removed, and the table is grouped
and aggregated, as shown in Figure 8-23. Grouped Rows will be
added to the Applied Steps list. I have also sorted the table by the
Make and CountryName columns to make the grouping easier to
comprehend.

	A^B_C Make	1↕	A^B_C CountryName	2↕	1.2 TotalSales	▼	1.2 AverageCost	▼
1	Aston Martin		France		1487210		37232.14286	
2	Aston Martin		Switzerland		543170		22394.23077	
3	Aston Martin		USA		3722160		64055	
4	Aston Martin		United Kingdom		4933500		84149.47368	
5	Bentley		France		394250		46350	
6	Bentley		Spain		46750		30700	
7	Bentley		Switzerland		222750		28200	
8	Bentley		USA		1879750		46462.96296	
9	Bentley		United Kingdom		2407750		43685.71429	
10	Jaguar		France		212000		30500	
11	Jaguar		Germany		74750		42500	
12	Jaguar		Spain		69250		37500	
13	Jaguar		Switzerland		450250		34285.71429	
14	Jaguar		USA		2751000		39036.53846	
15	Jaguar		United Kingdom		2761750		37988.52459	
16	MGB		USA		348000		11166.66667	

Figure 8-23. *Grouping a dataset*

If you have created a really complex group and then realized that you need to change the order of the columns, all is not lost. You can alter the order of the columns in the output by clicking the ellipses to the right of each column definition in the Group By dialog and selecting Move Up or Move Down. The order of the columns in the Group By dialog will be the order of the columns (left to right) in the resulting dataset.

If you have defined a column group and then wish to modify it later, you can modify the contents of the "Group By" dialogue by clicking the settings cog on the corresponding step in the Applied Steps list.

Note   You do not have to Ctrl-click to select the grouping columns. You can add them one by one to the Group By dialog by clicking the Add Aggregation button (which may mean switching to the Advanced tab in the Group By dialogue). Equally, you can remove grouping columns (or added and aggregated columns) by clicking the ellipses to the right of a column name and selecting Delete from the pop-up menu.

Saving Changes in Power Query

As you would expect, you can save any changes that you have made when using Power Query at any time. However, you need to be aware that when you click the Save icon above the ribbon (or if you click File ➤ Save), you will be presented with a choice in the dialog that you can see in Figure 8-24.

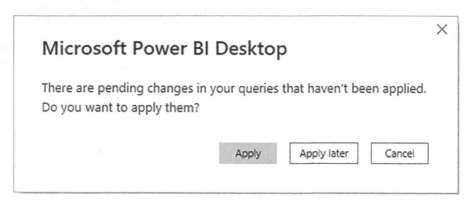

Figure 8-24. *Applying pending changes before saving in Power Query*

At this point, you have to decide if you want only to save the work that you have done using Power Query, or if you want not only to save your work but also update the data in the data model with the latest version of the data that results from the data transformation that you have carried out.

Consequently, you have to choose between

- *Apply*: Apply the changes and load the data to the data model.

- *Apply later*: Save your modifications but leave the data as it is currently—and potentially update the data later.

Exiting Power Query

In a similar vein to the Save options just described, you can choose how to exit Power Query and return to your reports and dashboards in Power BI Desktop. The default option (when you click the Close & Apply button) is to apply all the changes that you have made to the data, update the data model with the new data, and return to Power BI Desktop.

However, you have two other options that may prove useful. These appear in the menu for the Close & Apply button:

- *Apply*: Apply the changes that you have made to the data model while remaining in Power Query. This may involve changes to the fields in the Fields list.

- *Close*: Close Power Query but do not apply any changes.

If you choose not to apply the changes that you have made, then you will see the alert that is displayed in Figure 8-25. At some point, you will have to update the data model to ensure that you are using the correct data for your reports by clicking the Apply Changes button in the alert that appears above the report canvas.

Figure 8-25. *The pending changes alert*

Conclusion

This chapter started you on the road to transforming datasets with Power BI Desktop. You saw how to trim datasets by removing rows and columns. You also learned how to subset a sample of data from a data source by selecting alternating groups of rows.

You also saw how to choose the columns that you want to use in reports, how to move columns around in the dataset, and how to rename columns so that your data is easily comprehensible when you use it later in dashboards and reports. Then, you saw how to filter and sort data, as well as how to remove duplicates to ensure that your dataset only contains the precise rows that you need for your upcoming visualizations. Finally, you learned how to group and aggregate data.

It has to be admitted, nonetheless, that preparing raw data for use in dashboards and reports is not always easy and can take a while to get right. However, Power Query can make this task really fast and fluid with a little practice. So now that you have grasped the basics, it is time to move on and discover some further data transformation techniques. Specifically, you will see how to shape and further transform the data that you have imported. This is the subject of the next chapter.

CHAPTER 9

Shaping Data

In the previous chapter, you saw how to hone your dataset so that you defined only the rows and columns of data that you really need. Then you learned how to carry out basic transformations on the data that they contain. In this chapter, you will learn how to build on these foundations to deliver data that is ready to be molded into a structured and usable data model.

The generic term for this kind of data preparation in Power BI Desktop is *shaping data*. It covers the following:

- *Joining queries*: This involves taking two queries and linking them so that you display the data from both sources as a single dataset. You will learn how to extend a query with multiple columns from a second query as well as how to aggregate the data from a second query and add this to the initial dataset. You will also see how to create complex joins when merging queries.

- *Pivoting and unpivoting data*: If you need to switch data in rows to display as columns—or vice versa—then you can get Power Query to help you do exactly this. This means that you can guarantee that the data in all the tables that you are using conforms to a standardized tabular structure that is essential for Power BI to function efficiently.

- *Transposing data*: This can be required to switch columns into rows and vice versa.

- *Data quality analysis*: This means using Power Query to give you an idea about the quality of the available data.

These techniques can be—and probably will be—used alongside many of the techniques that you saw previously in Chapters 7 and 8. After all, one of the great strengths of Power Query is that it recognizes that data transformation is a complex

© Adam Aspin 2022
A. Aspin, *Pro Data Mashup for Power BI*, https://doi.org/10.1007/978-1-4842-8578-7_9

business and consequently does not impose any strict way of working. Indeed, it lets you experiment freely with a multitude of data transformation options. So remember that you are at liberty to take any approach you want when transforming source data. The only thing that matters is that it gives you the result that you want.

Merging Data

Until now, we have treated each individual query as if it existed in isolation. The reality, of course, is that you will frequently be required to use the output of one query in conjunction with the output of another to join data from different sources in various ways. Assuming that the results of one query share a common field (or fields) with another query, you can "join" queries into a single "flattened" data table. Power BI Desktop calls this a merge operation, and it enables you, among other things, to

- Look up data elements in another "reference" table to add lookup data. For example, you may want to add a client name where only the client reference exists in your main table.

- Aggregate data from a "detail" table (such as invoice lines) and include the totals in a higher-grained table, such as a table of invoices.

Here, again, the process is not difficult. The only fundamental factor is that the two tables, or queries, that you are merging must have a shared field or fields that enable the two tables to match records coherently. Indeed, the field headers/titles don't need to be exactly the same to be able to perform the join. Let's look at a couple of examples.

Adding Data

First, let's try looking up extra data that we will add to a query:

1. In a new, empty Power BI Desktop file, load both the worksheets (clients and Sales) in the `C:\PowerBiDesktopSamples\SalesData.xlsx` Excel file.

2. Click the Transform Data button in the Home ribbon.

3. Click the query named Sales in the Queries pane of the Power Query window.

4. Click the Merge Queries button in the Home ribbon. The Merge dialog will appear.

5. In the upper part of the dialog—where an overview of the output from the current query is displayed—scroll to the right and click the ClientName column title. This column is highlighted.

6. In the drop-down menu under the upper table, select the Clients query. The output from this query will appear in the lower part of the dialog.

7. In the lower table, select the column title for the column—the join column—that maps to the column that you selected in step 5. This will also be the ClientName column. This column is then selected in the lower table. You may be asked to set privacy levels for the data sources. If this is the case, set them to Public.

8. Select Inner (only matching rows) from the Join Kind pop-up menu. The dialog will look like Figure 9-1.

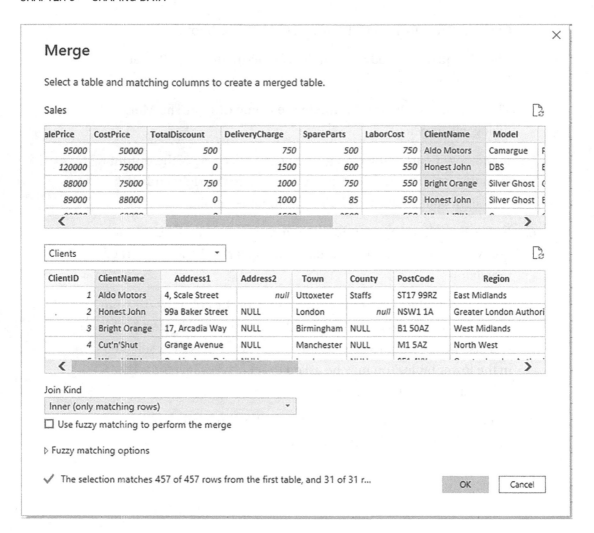

Figure 9-1. *The Merge dialog*

9. Click OK. A new column is added to the right of the existing data table. It is named Clients—representing the merged table. Merged Queries is added to the Applied Steps list.

10. Scroll to the right of the existing data table. The new column contains the word Table in every cell. This column will look something like Figure 9-2.

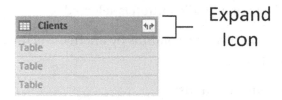

Figure 9-2. *A new, merged column*

11. Click the Expand icon to the right of the added column name (it should have the name of the second query that you merged into a source query). The pop-up list of all the available fields in this data table (or query, if you prefer) is displayed, as shown in Figure 9-3.

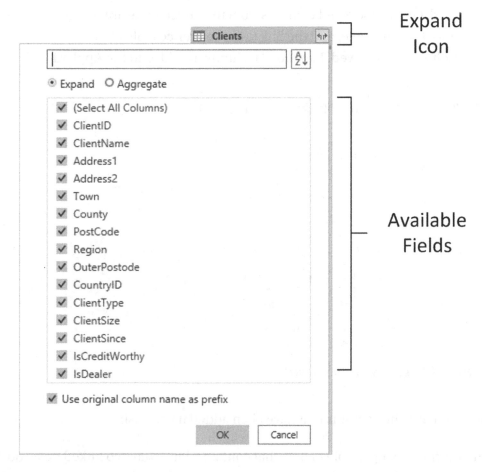

Figure 9-3. *The fields available in a joined query*

12.   Ensure that the Expand radio button is selected.

13.   Clear the selection of all the columns by unchecking the (Select All Columns) check box.

14.   Select the following columns:

   a. *ClientName*: Note that this will result in what's effectively a duplicate column as it also appears in the "parent" table.

   b. *ClientSize*

   c. *ClientSince*

15.   Uncheck Use original column name as prefix.

16.   Click OK. The selected columns from the linked table are merged into the main table, and the link to the reference table (New Column) is removed. Expanded Columns is added to the Applied Steps list.

17.   The result should look like that in Figure 9-4.

A͟B͟C VehicleType	A͟B͟C InvoiceNumber	A͟B͟C ClientName.1	A͟B͟C ClientSize	ABC₁₂₃ ClientSince
Saloon	8B3D7F83-F42C-4523-A737-CDCBF7705...	Aldo Motors	Large	04/01/1998 00:00:00
Coupe	15A38C61-82BD-4CCD-8F0B-49EEE59AF...	Aldo Motors	Large	04/01/1998 00:00:00
Coupe	CFC6726D-1522-4981-BC6B-766AE2C6E...	Aldo Motors	Large	04/01/1998 00:00:00
Coupe	CFC6726D-1522-4981-BC6B-766AE2C6E...	Aldo Motors	Large	04/01/1998 00:00:00
Saloon	A44A6460-1BBF-4AFB-A9F8-248680274...	Aldo Motors	Large	04/01/1998 00:00:00
Saloon	00B971F3-33DD-4DB0-B08E-973E4B531...	Aldo Motors	Large	04/01/1998 00:00:00
Coupe	139BEEEF-FF32-4BE9-9EF1-819AC888B8...	Honest John	Large	01/01/2000 00:00:00
Saloon	2ABAA300-E2A5-4E37-BFCA-7B80ED88A...	Honest John	Large	01/01/2000 00:00:00
Saloon	EA924C34-F8FD-4E2F-9334-960E90BA61...	Honest John	Large	01/01/2000 00:00:00
Saloon	A5FC5D6E-D875-47E5-8F4B-B5EC81867...	Honest John	Large	01/01/2000 00:00:00
Saloon	D35D72CD-5FF3-4701-A6D1-265A4F4E7...	Bright Orange	Large	01/04/2005 00:00:00
Coupe	F1B566F0-D137-4810-B449-575438F3F3...	Bright Orange	Large	01/04/2005 00:00:00
Coupe	B47CA156-7077-4690-AA1A-0CF4519BA...	Bright Orange	Large	01/04/2005 00:00:00
Saloon	C6ECDC8D-5356-4E10-B06C-DF176E7A4...	Bright Orange	Large	01/04/2005 00:00:00
Saloon	D63E42B5-4282-4DE8-9A68-F89E8B0C7...	Bright Orange	Large	01/04/2005 00:00:00

Figure 9-4. *Merged column output*

18.   Rename the columns that have been added if you wish.

You now have a single table of data that contains data from two linked data sources. Refreshing the Sales query will also reprocess the dependent *clients* query and result in the latest version of the data being reloaded into the merged query.

Note   You probably noticed that the Merge dialog indicated how many matching records there were in the two queries. This can be a useful indication that you have selected the correct column(s) to join the two queries.

Aggregating Data During a Merge Operation

If you are not just looking up reference data but need to aggregate data from a separate table and then add the results to the current query, then the process is largely similar. This second approach, however, is designed to suit another completely different requirement. Previously, you saw the case where the current query had many records that mapped to a *single* record in the lookup table. This second approach is for when your current (or main) query has a single record where there are *multiple* linked records in the second query. Consequently, you need to aggregate the data in the second table to bring the data across into the first table. Here is a simple example, using some of the sample data from the `C:\PowerBiDesktopSamples\` folder:

1. In a new Power BI Desktop file, load only the tables Invoices and InvoiceLines from the InvoicesAndInvoiceLines.xlsx Excel source file in the `C:\PowerBiDesktopSamples` folder. Click Transform Data to see the two worksheets it contains (Invoices and InvoiceLines) in Power Query. This will create two queries.

2. Click the query named Invoices in the Queries pane on the left.

3. In the Home ribbon, click the Merge Queries button. The Merge dialog will open. You will see some of the data from the Invoices dataset in the upper part of the dialog.

4. Click anywhere inside the InvoiceID column. This column is selected.

5. In the popup, select the InvoiceLines query. You will see some of the data from the InvoiceLines dataset in the lower part of the dialog.

6. Click anywhere inside the InvoiceID column for the lower table. This column is selected.

7. Select Inner (only matching rows) from the Join Kind pop-up menu. The dialog will look like Figure 9-5.

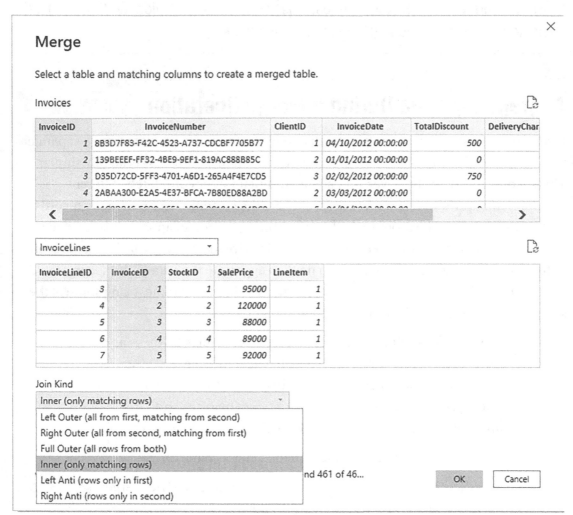

Figure 9-5. *The Merge dialog when aggregating data*

8. Click OK. The Merge dialog will close, and a new column named InvoiceLines will be added at the right of the Invoices query. Merged Queries is added to the Applied Steps list.

9. Scroll to the right of the existing data table. You will see the new column (named InvoiceLines) that contains the word *Table* in every cell.

10. Click the Expand icon to the right of the new column title (the two arrows facing left and right). The pop-up list of all the available fields in the InvoiceLines query is displayed.

11. Select the Aggregate radio button.

12. Select the Sum of SalePrice field and uncheck all the others.

13. Uncheck the "Use original column name as prefix" check box. The dialog will look like Figure 9-6.

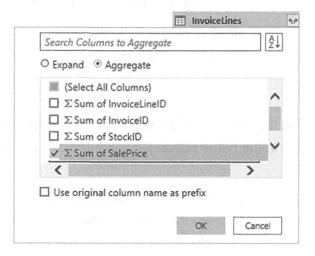

Figure 9-6. *The available fields from a merged dataset*

14. Click OK. Aggregated InvoiceLines is added to the Applied Steps list.

Power BI Desktop will add up the total sale price for each invoice and add this as a new column. Naturally, you can choose the type of aggregation that you wish to apply (before clicking OK), if the sum is not what you want. To do this, place the cursor over the column that you want to aggregate (see step 11 in the preceding exercise) and click the drop-down menu at the right of the field name. Power BI Desktop will suggest a set of options. The available aggregation options are explained in Table 9-1.

Table 9-1. *Merge Aggregation Options*

Option	Description
Sum	Returns the total value of the field
Average	Returns the average value of the field
Median	Returns the median value of the field
Minimum	Returns the minimum value of the field
Maximum	Returns the maximum value of the field
Count (All)	Counts all records in the dataset
Count (Not Blank)	Counts all records in the dataset that are not empty

Tip   If you loaded the data instead of editing the query in step 1, simply click the Transform Data button in the Home ribbon to switch to Power Query.

The merge process that you have just seen, while not complex in itself, suddenly opens up many new horizons. It means that you can now create multiple separate queries that you can then use together to expand your data in ways that allow you to prepare quite complex datasets.

Here are a couple of comments I need to make about the merge operation:

- Only queries that have been previously created in Power Query can be used when merging datasets. So remember to connect to all the datasets that you require before attempting a merge operation.

- Refreshing a query will cause any other queries that are merged into this query to be refreshed also. This way, you will always get the most up-to-date data from all the queries in the process.

Note   You can also merge a parent table and a child table *without* aggregating. This will add any fields from the child table to the fields from the parent table exactly as you saw in the first merge example. However, this is (probably) a "one-to-many" join. It means that data from the parent table will be duplicated in the rows where there is more than one record in the child table.

Types of Join

When merging queries—either to join data or to aggregate values—you are faced with a choice when it comes to how to link the two queries. The choice of join can have a profound effect on the resulting dataset. Consequently, it is important to understand the six join types that are available. These are described in Table 9-2.

Table 9-2. *Join Types*

Join Type	Explanation
Left Outer	Keeps all records in the upper dataset in the Merge dialog (the dataset that was active when you began the merge operation). Any matching rows (those that share common values in the join columns) from the second dataset are kept. All other rows from the second dataset are discarded.
Right Outer	Keeps all records in the lower dataset in the Merge dialog (the dataset that was not active when you began the merge operation). Any matching rows (those that share common values in the join columns) from the upper dataset are kept. All other rows from the upper dataset (the dataset that was active when you began the merge operation) are discarded.
Full Outer	All rows from both queries are retained in the resulting dataset. Any records that do not share common values in the join field(s) contain blanks in certain columns.
Inner	Only joins queries where there is an exact match on the column(s) that are selected for the join. Any rows from either query that do not share common values in the join column(s) are discarded.
Left Anti	Keeps only rows from the upper (first) query that don't have any matching rows from the lower (second) query.
Right Anti	Keeps only rows from the lower (second) query that don't have any matching rows from the upper (first) query.

Note   When you use any of the *outer* joins, you are keeping records that do not have any corresponding records in the second query. Consequently, the resulting dataset contains empty values for some of the columns.

When you are expanding the column that is the link to a merged dataset, you have a couple of useful options that are worth knowing about:

- Use original column name as prefix

- Search columns to expand

Use the Original Column Name As the Prefix

You will probably find that some columns from joined queries can have the same names in both source datasets. It follows that you need to identify which column came from which dataset. If you leave the check box selected for the "Use original column name as prefix" merge option (which is the default), any merged columns will include the source query name to help you identify the data more accurately.

If you find that these longer column names only get in the way, you can unselect this check box. This will leave the added columns from the second query with their original names. However, because Power BI Desktop cannot accept duplicate column names, any new columns will have .1, .2, and so forth, added to the column name.

Search Columns to Expand

If you are merging a query with a second query that contains a large number of columns, then it can be laborious to scroll down to locate the columns that you want to include. To narrow your search, you can enter a few characters from the name of the column that you are looking for in the Search columns to aggregate box (or Search Columns to Expand if you're expanding rather than aggregating). The more characters you type, the fewer matching columns are displayed in the Expansion pop-up dialog.

Joining on Multiple Columns

In the examples so far, you only joined queries on a single column. While this may be possible if you are looking at data that comes from a clearly structured source (such as a relational database), you may need to extend the principle when joining queries from diverse sources. Fortunately, Power BI Desktop allows you to join queries on *multiple* columns when the need arises.

As an example of this, the sample data contains a file that I have prepared as an example of how to join queries on more than one column. This sample file contains data from the sources that you saw in previous chapters. However, they have been modeled as a data warehouse star schema. To complete the model, you need to join a dimension named Geography to a fact table named Sales so that you can add the field GeographySK to the fact table. However, the Sales table and the Geography table share two fields (Country and Region) that must correspond for the queries to be joined. The following explains how to perform a join using multiple fields:

1. In a new Power BI Desktop file, click Get Data, select Excel as the source, and in the `C:\PowerBiDesktopSamples\StarSchema.xlsx` Excel file, select the two worksheets (Geography and Sales).

2. Click Transform Data to open the Power Query. Alternatively, if you are already in Power Query, simply click New Source, and select Excel as the source data type and load the two worksheets directly.

3. Select the Sales query from the list of existing queries from the Queries pane on the left of the Power Query window.

4. In the Home ribbon, click the Merge Queries button. The Merge dialog will appear.

5. In the pop-up list of queries, select Geography as the second query to join to the first (upper) query.

6. Select Inner (only matching rows) from the Join Kind popup.

7. In the upper list of fields (taken from the Sales table), Ctrl-click the fields CountryName and Region, *in this order*. A small number will appear to the right of each column header, indicating the order that you selected the columns.

8. In the lower list of fields (taken from the Sales table), Ctrl-click the fields CountryName and Region, *in this order*. A small number will appear to the right of each column header, indicating the order that you selected the columns.

9. Verify that you have a reasonable number of matching rows in the information message at the bottom of the dialog. The dialog will look like Figure 9-7.

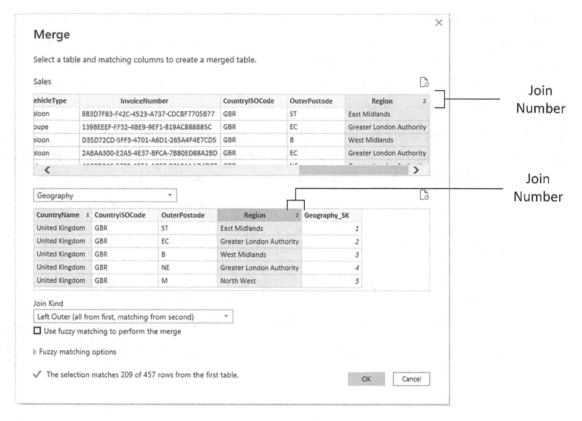

Figure 9-7. Joining queries using multiple columns

10.   Click OK. Merged Queries is added to the Applied Steps list.

You can then continue restructuring your data. In this example, that would be adding the GeographySK field to the Sales query and then removing the Country and Region fields from the Sales query.

There is no real limit to the number of columns that can be used when joining queries. It will depend entirely on the shape of the source data. However, each column used to define the join must exist in both datasets, and each pair of columns must be of the same (or a similar) data type.

Fuzzy Matching

Fuzzy matching allows you to compare values in the columns to join in a merge operation by using fuzzy matching logic, rather than looking for exact matches in the data.

1. Start a merge operation (as you did previously) and select the two tables to merge as well as the column(s) used to join the queries.

2. In the Merge dialog, check Use fuzzy matching to perform the merge.

3. Set a similarity threshold. This defines the "latitude" that Power Query can apply when looking for similar elements to match across the two queries.

4. Enter a figure for the maximum number of matches. The dialog will look like Figure 9-8.

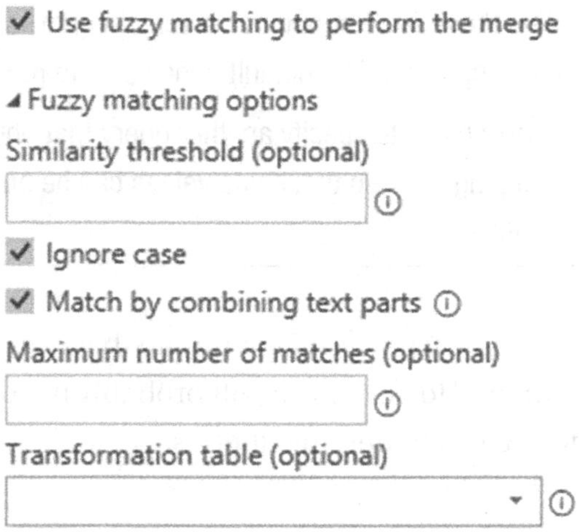

☑ Use fuzzy matching to perform the merge

◢ Fuzzy matching options
Similarity threshold (optional)
[   ] ⓘ
☑ Ignore case
☑ Match by combining text parts ⓘ
Maximum number of matches (optional)
[   ] ⓘ
Transformation table (optional)
[▾] ⓘ

Figure 9-8. *Fuzzy matching*

Fuzzy Matching Options

There are a few fuzzy matching options that you might want to apply—and that you certainly need to understand when you use this technique. They are explained in Table 9-3.

Table 9-3. *Fuzzy Matching Options*

Option	Explanation
Similarity threshold	Indicates how similar two values need to be in order to match. The minimum value of 0.00 will cause all values to match each other, and the maximum value of 1.00 will only allow exact matches. The default is 0.80.
Ignore case	Ignores uppercase and lowercase characters in the two columns.
Match by combining text parts	Joins words to use for matching
Maximum number of matches	Sets the maximum number of matching rows that will be returned for each input row. The default behavior is to return all matches.
Transformation table	Allows users to specify another query that holds a manually prepared mapping table so that some values can be auto-mapped as part of the matching.

Fuzzy matching is an extremely powerful technique that can assist you in joining datasets from disparate sources. However, you will probably need to experiment to find the appropriate options to use when merging queries.

Merge As New Query

In the preceding merge process, you merged data into a source table. However, there may be occasions when you prefer to keep the source query (or table, if you prefer) intact and output the result of the merge operation into a new query.

To do this, the trick is to click not the Merge button itself, but the arrow to the right of the Merge button—and selecting Merge Queries as New. You can see this option in Figure 9-9.

Figure 9-9. *Merging tables as a new query*

This will create a new query (named Merge1—but you can rename it later) and leave the initial queries intact.

Preparing Datasets for Joins

You could have to carry out a little preparatory work on real-world datasets before joining queries. More specifically, any columns that you join have to be the same basic data type. Put simply, you need to join text-based columns to other text-based columns, number columns to number columns, and date columns to date columns. If the columns are *not* the same data type, you receive a warning message when you try to join the columns in the Merge dialog.

Consequently, it is nearly always a good idea to take a look at the columns that you will use to join queries *before* you start the merge operation itself. Remember that data types do not have to be identical, just similar. So a decimal number type can map to a whole number, for instance.

You might also have to cleanse the data in the columns that are used for joins before attempting to merge queries. This could involve the following:

- Removing trailing or leading spaces in text-based columns

- Isolating part of a column (either in the original column or as a new column) to use in a join

- Verifying that appropriate data types are used in join columns

Correct and Incorrect Joins

Merging queries is the one data restructuring operation that is often easier in theory than in practice, unfortunately. If the source queries were based on tables in a relational or even dimensional database, then joining them could be relatively easy, as a data architect will (hopefully) have designed the database tables to allow for them to be joined. However, if you are joining two completely independent queries, then you could face several major issues:

- The columns do not map.

- The columns map, but the result is a massive table with duplicate records.

Let's take a look at these possible problems.

The Columns Do Not Map

If the columns do not map (i.e., you have joined the data but get no resulting records), then you need to take a close look at the data in the columns that you are using to establish the join. The questions you need to ask are as follows:

- Are the values in the two queries the same data type?

- Do the values really map—or are they different?

- Are you using the correct columns?

- Are you using too many columns and so specifying data that is not in both queries?

The Columns Map, but the Result Is a Massive Table with Duplicate Records

Joining queries depends on isolating *unique* data in both source queries. Sometimes, a single column does not contain enough information to establish a unique reference that can uniquely identify a row in the query.

In these cases, you need to use two or more columns to join queries—or else rows will be duplicated in the result. Therefore, once again, you need to look carefully at the data and decide on the minimum number of columns that you can use to join queries correctly.

Examining Joined Data

Joining data tables is not always easy. Neither is deciding if the outcome of a merge operation will produce the result that you expect. So Power Query includes a solution to these kinds of dilemma. It can help you more clearly see what a join has done. More specifically, it can show you for each record in the first query exactly which rows are joined from the second query.

Do the following to see this in action:

1. Carry out steps 1 through 10 in the example you saw earlier ("Joining on Multiple Columns" section).

2. Scroll to the right in the data table. You will see the new column named Geography (as shown in Figure 9-10).

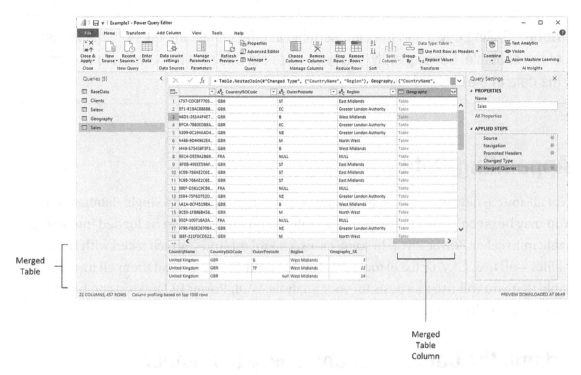

Merged
Table

Merged
Table
Column

Figure 9-10. *Joined data*

3. Click to the *right* of the word *Table* in the row where you want to
    see the joined data. Note that you must *not* click the word *Table*.
    A second table will appear under the main query's data table
    containing the data from the second query that is joined for this
    particular row. Figure 9-11 shows an example of this.

This technique is as simple as it is useful. There are nonetheless a few comments that
I need to make:

*   You can resize the lower table (and consequently display more or less
    data from the second joined table) by dragging the bottom border of
    the top data table up or down.

*   Clicking the word *Table* in the Geography column adds a new step
    to the query that replaces the source data with the linked data. You
    can also do this by right-clicking inside the NewColumn column and
    selecting Drill Down.

Note   Drilling down into the merged table in effect limits the query to the row(s) of the subtable. Consequently, you have to delete this step if you want to access all the data in the merged tables.

Appending Data

Not all source data is delivered in its entirety in a single file or as a single database table. You may be given access to two or more tables or files that have to be loaded into a single table in Excel or Power Pivot. In some cases, you might find yourself faced with hundreds of files—all text, CSV, or Excel format—and the requirement to load them all into a single table that you will use as a basis for your analysis. Well, Power BI Desktop can handle these eventualities too.

Adding the Contents of One Query to Another

In the simplest case, you could have two data sources that are structurally identical (i.e., they have the same columns in the same order), and all that you have to do is add one to another to end up with a query that outputs the amalgamated content of the two sources. This is called *appending data*, and it is easy, provided that the two data sources have *identical* structures; this means

- They have the same number of columns.

- The columns are in the same order.

- The data types are identical for each column.

- The columns have the same names.

As long as all these conditions are met, you can append the output of queries (which Power BI Desktop also calls *Tables* and many people, including me, refer to as datasets) one into another. The queries do not have to have data that comes from identical source types, so you can append the output from a CSV file to data that comes from an Oracle database, for instance. As an example, we will take two text files and use them to create one single output:

1. Create queries to load each of the following text files into Power BI Desktop—without the final load step, which would add them to the data model. Both files are in the `C:\PowerBiDesktopSamples\MultipleIdenticalFiles` folder:

    a. Colours_01.txt

    b. Colours_02.txt

2. You can see the contents of these two queries in Figure 9-11.

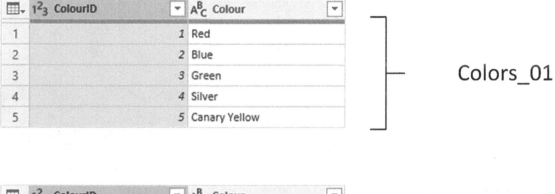

Figure 9-11. *Source data for appending*

3. Select one of the queries (I use Colours_01, but either will do).

4. Click the arrow to the right of the Append Queries button in the Power Query Home ribbon. You can see this in Figure 9-12.

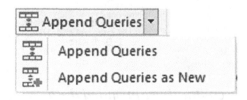

Figure 9-12. *The Append menu choices*

5. Select Append queries as new. The Append dialog will appear.

6. Ensure that the Two tables radio button is selected.

7. From the Second table popup, choose the query Colours_02. The dialog will look like the one in Figure 9-13.

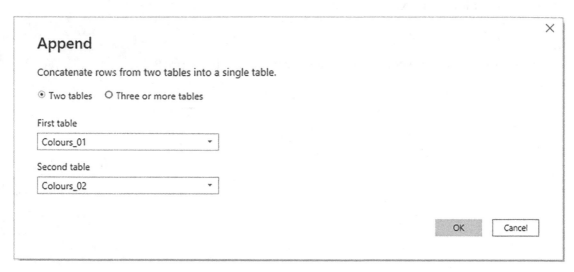

Figure 9-13. *The Append dialog*

8. Click OK. The data from the two output tables is appended in a new query. You can see an example of the resulting output in Figure 9-14.

123 ColourID	ABC Colour
1	Red
2	Blue
3	Green
4	Silver
5	Canary Yellow
6	Night Blue
7	Black
8	British Racing Green
9	Dark Purple
10	Pink

Figure 9-14. *A new query containing appended data*

You can now continue with any modifications that you need to apply. You will notice that the column names are not repeated as part of the data when the tables are appended one to the other.

The new query is named Append1. You can, of course, rename this to something more comprehensible.

One interesting aspect of this approach is that you have created a link between the two source tables and the new query. This means that when you refresh the source data, not only are the data in the tables Colours_01 and Colours_02 updated, but the "derived" query that you just created is updated as well.

Appending the Contents of Multiple Queries

Power Query does not limit you to appending only two files at once. You can (if you really need to) append a virtually limitless number of identical files.

Moreover, you can append Excel files just as easily as you can append text or CSV files—as the following example shows:

1. In a new, empty Power BI Desktop file, click Transform Data to open Power Query.

2. Create queries to load all of the Excel files in the folder `C:\PowerBiDesktopSamples\MultipleIdenticalExcel` (the files are identical, and each one contains a single worksheet named BaseData).

3. Select the query named BaseData in the Queries pane on the left.

4. Click the Append Queries button.

5. Select the Three or more tables radio button in the Append dialog.

6. Ctrl-click the tables "BaseData (2)" and "BaseData (3)" in the Available table(s) list on the left of the dialog.

7. Click the Add button. You can see what the Append dialog now looks like in Figure 9-15.

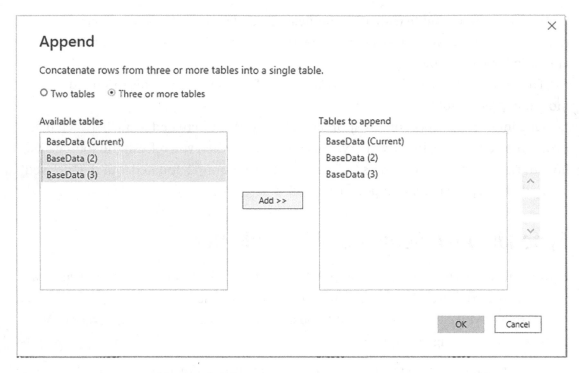

Figure 9-15. *Appending multiple queries*

8.   Click OK. The data from the queries "BaseData (2)" and "BaseData (3)" will be appended to the current query (BaseData). Appended Query will be added to the Applied Steps list.

It is worth noting the following:

- You can remove queries from the list of queries to append on the right by clicking the query (or Ctrl-clicking multiple queries) and subsequently clicking the cross icon on the right of the dialog.

- You can alter the load order of queries by clicking the query to move and then clicking the up and down chevrons on the right of the dialog.

Note   Appending queries presumes that the source data has already been loaded into Power Query. This is not the same as loading multiple files from the same directory—which is explained in Chapter 12.

Changing the Data Structure

Sometimes, your requirements go beyond the techniques that we have seen so far when discussing data cleansing and transformation. Some data structures need more radical reworking, given the shape of the data that you have acquired. I include in this category the following:

- Unpivoting data

- Pivoting data

- Transforming rows and columns

Each of these techniques is designed to meet a specific, yet frequent, need in data shaping, and all are described in the next few pages.

Unpivoting Tables

From time to time, you may need to analyze data that has been delivered in a "pivoted" or "denormalized" format. Essentially, this means that information that really should be in a single column has been broken down and placed across several columns. An example of the first few rows of a pivoted dataset is given in Figure 9-16 and can be found in the C:\PowerBiDesktopSamples\PivotedDataSet.xlsx sample file.

	A	B	C	D	E	F	G	H
1	InvoiceDate ▼	Aston Martin ▼	Bentley ▼	Jaguar ▼	MGB ▼	Rolls Royce ▼	Triumph ▼	TVR ▼
2	02/01/2013	75890	25700	88200	4500	62000	8500	
3	09/01/2013	31125						
4	10/01/2013	17500						
5	02/02/2013	75890	25700	63200	8500	62000	17000	37500
6	11/02/2013	22500						
7	02/03/2013	75890	25700	88200	4500	75890	8500	
8	12/03/2013	17500						
9	13/03/2013					31125		
10	14/03/2013	17500						
11	02/04/2013	75890	25700	99500	8500	62000	17000	37500
12	15/04/2013					22500		
13	16/04/2013	17500						
14	02/05/2013	75890	62000	124500	4500	75890	8500	
15	17/05/2013	17500						
16	18/05/2013	17500						
17	19/05/2013	22500						
18	02/06/2013	62000	62000	63200	8500	62000	17000	37500
19	20/06/2013	17500						
20	02/07/2013	62000	25700	88200	4500	62000	17000	
21	21/07/2013					17500		
22	22/07/2013	22500						
23	02/08/2013	62000	62000	38200	8500	62000	17000	37500
24	02/09/2013	62000	62000	124500	4500	75890	17000	
25	23/09/2013	17500						
26	02/10/2013	62000	62000	63200	8500	75890	17000	37500
27	24/10/2013					17500		
28	02/11/2013	125000	25700	87000	4500	75890	17000	37500
29	25/11/2013	31125						
30	26/11/2013	17500						
31	27/11/2013	17500						
32	02/12/2013	125000	25700	137000	4500	62000	17000	

Figure 9-16. *A pivoted dataset*

To analyze this data correctly, we really need the makes of the cars to be switched from being column titles to becoming the contents of a specific column. Fortunately, this is not hard at all:

1. In a new Power BI Desktop file, load the PivotedCosts worksheet from the C:\PowerBiDesktopSamples\PivotedDataSet.xlsx file into Power BI Desktop. Ensure that the first row is set to be the table headers (this should happen automatically and is indicated by the Promoted Headers step in the Applied Steps pane).

2. Switch to Power Query (unless it is already open) and select all the columns that you want to unpivot. In this example, this means all columns except the first one (all the makes of cars).

3. In the Transform ribbon, click the Unpivot Columns button (or right-click any of the selected columns and choose Unpivot Columns from the context menu). The table is reorganized, and the first few records look as they do in Figure 9-17. Unpivoted Columns is added to the Applied Steps list.

	InvoiceDate	A^B_C Attribute	1^2_3 Value
1	02/01/2013	Aston Martin	75890
2	02/01/2013	Bentley	25700
3	02/01/2013	Jaguar	88200
4	02/01/2013	MGB	4500
5	02/01/2013	Rolls Royce	62000
6	02/01/2013	Triumph	8500
7	09/01/2013	Aston Martin	31125
8	10/01/2013	Aston Martin	17500
9	02/02/2013	Aston Martin	75890
10	02/02/2013	Bentley	25700
11	02/02/2013	Jaguar	63200
12	02/02/2013	MGB	8500
13	02/02/2013	Rolls Royce	62000
14	02/02/2013	Triumph	17000
15	02/02/2013	TVR	37500
16	11/02/2013	Aston Martin	22500
17	02/03/2013	Aston Martin	75890
18	02/03/2013	Bentley	25700
19	02/03/2013	Jaguar	88200
20	02/03/2013	MGB	4500

Figure 9-17. An unpivoted dataset

4. Rename the columns that Power Query has named Attribute and Value.

The data is now presented in a standard tabular way, and so it can be used to create a data model and then produce reports and dashboards.

Unpivot Options

There are a couple of available options when you unpivot data using the Unpivot Columns button popup in the Transform ribbon:

- *Unpivot Other Columns*: This will add the contents of all the other (not selected) columns to the unpivoted output.

- *Unpivot Only Selected Columns*: This will only add the contents of any *preselected* columns to the unpivoted output.

Note   As is the case with so many of the techniques that you apply using Power Query, it is really important to select the appropriate column(s) before carrying out pivot and unpivot operations.

Pivoting Tables

On some occasions, you may have to switch data from columns to rows so that you can use it efficiently. This kind of operation is called *pivoting data*. It is—perhaps unsurprisingly—very similar to the unpivot process that you saw in the previous section.

1. Follow steps 1 through 3 of the previous section so that you end up with the table of data that you can see in Figure 9-17.

2. Click inside the column Attribute.

3. In the Transform ribbon, click the Pivot Column button. The Pivot Column dialog will appear.

4. Select Value (the column of figures) as the values column that is aggregated by the pivot transformation.

5. Expand Advanced options and ensure that Sum is selected as the Aggregate Value Function. The Pivot Column dialog will look like Figure 9-18.

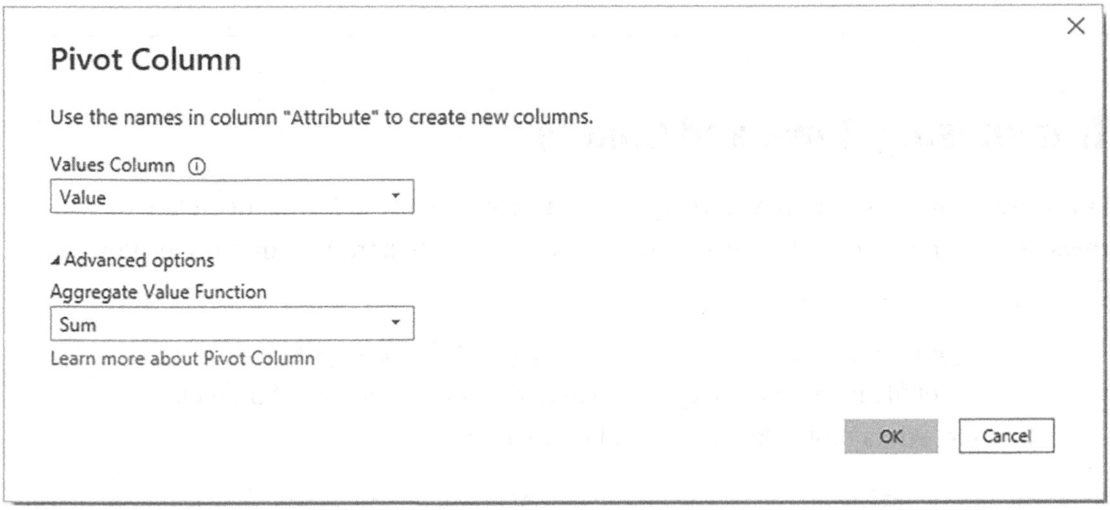

Figure 9-18. *The Pivot Column dialog*

6. Click OK. The table is pivoted and looks like Figure 9-19. Pivoted Column is added to the Applied Steps list.

	InvoiceDate	Aston Martin	Bentley	Jaguar	MGB	Rolls Royce	Triumph	TVR
1	02/01/2013	75890	25700	88200	4500	62000	8500	null
2	09/01/2013	31125	null	null	null	null	null	null
3	10/01/2013	17500	null	null	null	null	null	null
4	02/02/2013	75890	25700	63200	8500	62000	17000	37500
5	11/02/2013	22500	null	null	null	null	null	null
6	02/03/2013	75890	25700	88200	4500	75890	8500	null
7	12/03/2013	17500	null	null	null	null	null	null
8	13/03/2013	null	null	null	null	31125	null	null
9	14/03/2013	17500	null	null	null	null	null	null
10	02/04/2013	75890	25700	99500	8500	62000	17000	37500
11	15/04/2013	null	null	null	null	22500	null	null
12	16/04/2013	17500	null	null	null	null	null	null
13	02/05/2013	75890	62000	124500	4500	75890	8500	null
14	17/05/2013	17500	null	null	null	null	null	null
15	18/05/2013	17500	null	null	null	null	null	null
16	19/05/2013	22500	null	null	null	null	null	null
17	02/06/2013	62000	62000	63200	8500	62000	17000	37500
18	20/06/2013	17500	null	null	null	null	null	null
19	02/07/2013	62000	25700	88200	4500	62000	17000	null
20	21/07/2013	null	null	null	null	17500	null	null
21	22/07/2013	22500	null	null	null	null	null	null
22	02/08/2013	62000	62000	38200	8500	62000	17000	37500
23	02/09/2013	62000	62000	124500	4500	75890	17000	null
24	23/09/2013	17500	null	null	null	null	null	null
25	02/10/2013	62000	62000	63200	8500	75890	17000	37500
26	24/10/2013	null	null	null	null	17500	null	null
27	02/11/2013	125000	25700	87000	4500	75890	17000	37500
28	25/11/2013	31125	null	null	null	null	null	null
29	26/11/2013	17500	null	null	null	null	null	null
30	27/11/2013	17500	null	null	null	null	null	null
31	02/12/2013	125000	25700	137000	4500	62000	17000	null

Figure 9-19. *Pivoted data*

Note   The Advanced options section of the Pivot Column dialog lets you choose the aggregation operation that is applied to the values in the pivoted table. These were described in Table 9-1.

Transposing Rows and Columns

On some occasions, you may have a source table where the columns need to become rows and the rows columns. Fortunately, this is a one-click transformation for Power BI Desktop. Here is how to do it:

1. Open the `C:\PowerBiDesktopSamples\DataToTranspose.xlsx` Excel file in Power Query. You will need to select Sheet1. You will see a data table like the one in Figure 9-20.

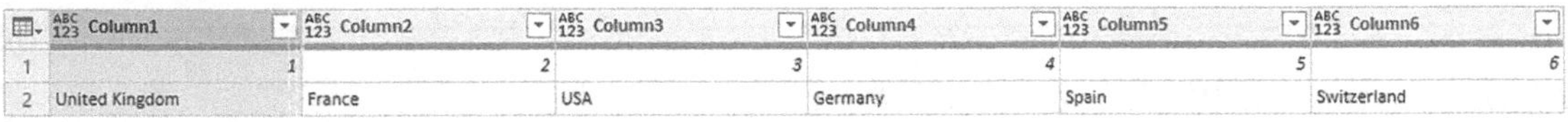

ABC 123 Column1	ABC 123 Column2	ABC 123 Column3	ABC 123 Column4	ABC 123 Column5	ABC 123 Column6
1	2	3	4	5	6
United Kingdom	France	USA	Germany	Spain	Switzerland

Figure 9-20.   *A dataset needing to be transposed*

2. In the Transform ribbon, click the Transpose button. The data is transposed and appears as two columns, just like the CountryList. txt file that you saw in Chapter 2. Transposed Table is added to the Applied Steps list.

3. Rename the resulting columns–as any headers in the original table will be lost and replaced with Column1, Column2, etc.

Data Quality Analysis

Power BI Desktop now includes the ability to help you analyze the quality and composition of your data. It can help you to get a deeper understanding and feel for your source data by providing information on

- Column quality

- Column distribution

- Column profile

Column Quality

Power BI Desktop can analyze the following aspects of every column of data in a query:

- Data validity

- Errors

- Empty fields

It then returns the percentages of each of these elements in each column. You can then examine the data in greater detail to track down potential anomalies. Here, we will use the file `Example1.pbix`.

1. In Power Query, switch to the View menu.

2. Check the Column quality check box. The column quality analysis will appear under each column header as you can see in Figure 9-21.

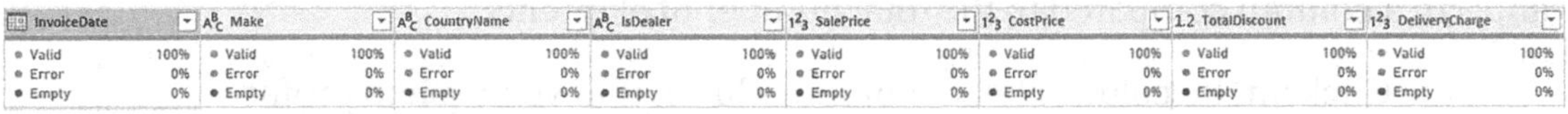

InvoiceDate		A^B_C Make		A^B_C CountryName		A^B_C IsDealer		1²₃ SalePrice		1²₃ CostPrice		1.2 TotalDiscount		1²₃ DeliveryCharge	
● Valid	100%	● Valid	100%	● Valid	100%	● Valid	100%	● Valid	100%	● Valid	100%	● Valid	100%	● Valid	100%
● Error	0%	● Error	0%	● Error	0%	● Error	0%	● Error	0%	● Error	0%	● Error	0%	● Error	0%
● Empty	0%	● Empty	0%	● Empty	0%	● Empty	0%	● Empty	0%	● Empty	0%	● Empty	0%	● Empty	0%

Figure 9-21. *Column quality*

Column Distribution

Analyzing the distribution of the data in a column shows unique and distinct values a column contains. This can help you to understand the relevance and usefulness of the data.

1. In the View menu, check the Column distribution check box. The column distribution analysis will appear under each column header as you can see in Figure 9-22.

Figure 9-22. Column distribution

It is worth noting the difference between distinct and unique elements in column distribution:

- *Distinct* elements may occur several times in a column—the total distinct count indicates how many separate elements exist overall.

- *Unique* elements appear only once in a column.

Column Profile

The column profile information indicates the relative percentages of each of the distinct values in a column compared to the total number of elements.

1. Click on the column header of the column that you wish to profile.

2. In the View menu, check the Column profile check box. The column profile analysis will appear at the bottom of the screen as shown in Figure 9-23.

Figure 9-23. Column profile

You may even choose to display all three data quality analysis options at the same time. If you do this, you could end up with a screen like the one shown in Figure 9-24.

Figure 9-24. *Displaying all data quality analysis options*

Of these options, only Column profile depends on the selected column. So you need to click each column independently to see the distribution of values for the selected column. Moreover, you can copy—independently—the column statistics and the value distribution for a column by clicking the ellipses at the top right of the column statistics and value distribution windows, respectively, and selecting Copy. This information can then be pasted into, for example, Excel to analyze the metadata for a dataset.

Right-clicking on the column profiling displays the Profiling context menu. This is shown in Figure 9-25.

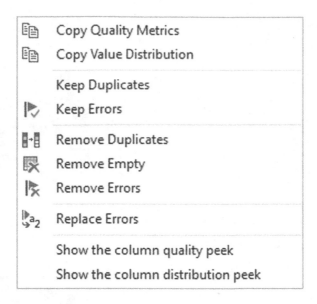

Figure 9-25. *Profiling context menu*

Profiling the Entire Dataset

By default, the built-in data profiling will only look at the first 1000 records in a dataset—presumably to speed up the profiling operation. However, you can force Power BI Desktop to profile the entire dataset.

1. At the bottom left of the Power Query screen, click Column profiling based on top 1000 rows. The popup that you can see in Figure 9-26 appears.

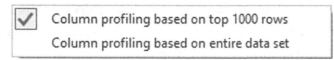

Figure 9-26. *Selecting the extent of the profiling*

2.   Select Column profiling based on entire data set. After a short
     delay (that will depend on the size of the dataset), the profiling
     analysis will be updated.

Correcting Anomalies

Power BI's data profiling tools only indicate what the data looks like from a high level.
It does not presume to modify the data—after all, the data might be useful even if it is
flagged as containing errors or anomalies.

However, it can help you to apply some basic data cleansing. Not only that, but a
lightbulb icon will appear if there are potential anomalies that you may want to correct.

1.   Hover over the column quality or column distribution data at the
     top of the screen. A popup containing more detailed information
     appears.

2.   Click the ellipses at the bottom right of the popup. The popup that
     you can see in Figure 9-27 appears.

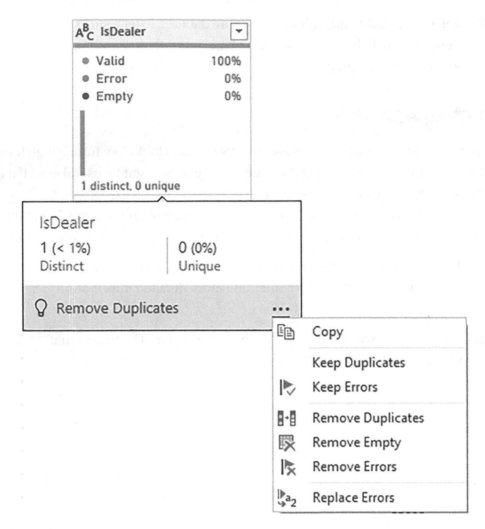

Figure 9-27. *Profiling correction options*

3. Select the option that interests you to correct any anomalies.

Of course, cleansing your data this way could actually damage the integrity of the data—and even remove valuable information. In some datasets, the data that is flagged as being an error could be the data that you want to examine in greater detail. So I advise caution before cleansing data simply to reduce the percentages of errors and empty cells to zero.

Note   Once you select any of the data quality options, they will remain active for future refreshes of this source data until you uncheck the option.

Data Transformation Approaches

I quite understand that you may be bewildered at the sheer number of available transformation options. So it may help, at this point, to remember a few key principles:

- If in doubt, right-click the column that you want to transform. This will list the most common available options in the context menu.

- To alter existing data, use the Transform menu.

- To add a new column, use the New Column menu. This is covered in greater detail in Chapter 11.

- Remember that you can "unwind" your modifications by deleting steps in the data transformation process.

Conclusion

This chapter showed you how to shape your source data into a valid data table from one or more potential sources. Among other things, you saw how to pivot and unpivot data, as well as how to transpose rows and columns.

Possibly the most important thing that you have learned is how to join individual queries so that you can add the data from one query into another. This can involve looking up data from a separate query or carrying the aggregated results from one query into another.

Finally, you learned how to analyze the quality of the source data and understand how immediately useful this dataset can be.

Now it is time to push your data transformation skills to the next level and learn how to cleanse raw source data. This is the subject of Chapter 10.

CHAPTER 10

Data Cleansing

Once a dataset has been shaped and structured (as covered in the previous two chapters), it probably still needs a good few modifications to make it ready for consumption. Many of these modifications are, at their heart, a selection of fairly simple yet necessary techniques that you apply to make the data cleaner and more standardized. I have chosen to group these approaches under the heading *data cleansing*.

The sort of things that you may be looking to do before finally loading source data into the data model normally cover a range of processes that *clean* the data so that it is usable in the data model. They can include the following:

- Change the data type for a column—by telling Power BI Desktop that the column contains numbers, for example.

- Ensure that the first row is used as headers (if this is required).

- Remove part of a column's contents.

- Replace the values in a cell with other values.

- Transform the column contents—by making the text uppercase, for instance, or by removing decimals from numbers.

This chapter will take you on a tour of these kinds of essential data cleansing operations. Once you have finished reading it, you should be confident that you can take a rough and ready data source as a starting point and convert it into a polished data table that is ready to become a pivotal part of your Power BI Desktop data model. Not only that, but you will have carried out really heavy lifting much faster and more easily (and much more cheaply) than you could have done using enterprise-level tools.

The sample data that you will need to follow the exercises in this chapter is in the folder `C:\PowerBiDesktopSamples`. Please note that all the examples in this chapter presume that you have opened Power Query.

© Adam Aspin 2022
A. Aspin, *Pro Data Mashup for Power BI*, https://doi.org/10.1007/978-1-4842-8578-7_10

Using the First Row As Headers

Power BI Desktop is very good at guessing if it needs to take the first record of a source dataset and have it function as the column headers. This is fundamental for two reasons:

- You avoid leaving the columns named Column1, Column2, and so on. Leaving them named generically like this would make it needlessly difficult for a user (or even yourself) to understand the data.

- You avoid having a text element (which should be the column title) in a column of figures, which can cause problems later on. This is because a whole column needs to have the *same* data type for that data type to be applied. Having a header text in the first row prevents this for numeric and date/time data types, for instance. This could be because the header is a text whereas the remainder of the column contains numbers or dates.

Yet there could be—albeit rare—occasions when Power BI Desktop guesses incorrectly and assumes that the first record in a dataset is data when it is really the header information. So instead of headers, you have a set of generic column titles such as Column1, Column2, and so forth. Fortunately, correcting this and using the first row as headers is a simple task:

1. Click Use First Row as Headers in the Home or Transform ribbon of the Power BI Desktop Query window.

After a few seconds, the first record disappears, and the column titles become the elements that were in the first record. The Applied Steps list on the right now contains a Promoted Headers element, indicating which process has taken place. This step is highlighted.

Note   Power BI Desktop is often able to apply this step automatically when the source is a database. It can often correctly guess when the source is a file. However, it cannot always guess accurately, so sometimes, you have to intervene. You can see if Power BI Desktop has had to guess this if it has added a Promoted Headers step to the Applied Steps list.

In the rare event that Power BI Desktop gets this operation wrong and presumes that a first row is column titles when it is not, you can reset the titles to be the first row by clicking the tiny triangle to the right of the Use First Row as Headers button. This displays a short menu where you can click the Use Headers as First Row option. The Applied Steps list on the right now contains a Demoted Headers element, and the column titles are Column1, Column2, and so forth. You can subsequently rename the columns as you see fit.

Changing Data Type

A truly fundamental aspect of data modification is ensuring that the data is of the appropriate type; that is, if you have a column of numbers that are to be calculated at some point, then the column should be a numeric column. If it contains dates, then it should be set to one of the date or datetime data types, and time data should be set to the time data type. I realize that this can seem arduous and even superfluous; however, *if you want to be sure that your data can be sliced and diced correctly further down the line*, then setting the right data types at the outset is *vital*. An added bonus is that if you validate the data types early on in the process of loading data, you can see from the start if the data has any potential issues—dates that cannot be read as dates, for instance. This allows you to decide what to do with poor or unreliable data early in your work with a dataset.

The good news here is that for many data sources, Power BI Desktop applies an appropriate data type. Specifically, if you have loaded data from a database, then Power BI Desktop will recognize the data type for each column and apply a suitable native data type. Unfortunately, things can get a little more painful with file sources, specifically CSV, text, and (occasionally) Excel files, as well as some XML files. In the case of these file types, Power BI Desktop often tries to guess the data type, but there are times when it does not succeed. If it has made a stab at deducing data types, then you see a Changed Type step in the Applied Steps list. Consequently, if you are obtaining your data from these sources, then you could well be obliged to apply data types to many of the columns manually.

Note   In some cases, numbers are not meant to be interpreted as numerical data. For instance, a French postal code is five numbers, but it will never be calculated in any way. So it is good practice to let Power BI Desktop know this by changing the data type to text in cases when a numeric data type is inappropriate.

Do the following to change data type for a column or a group of columns:

1. Open the Power BI Desktop sample file
   C:\PowerBiDesktopSamples\Example1.pbix.

2. Click the Transform Data button in the Home ribbon. Power
   Query will open.

3. Click inside the column whose data type you wish to change. If
   you want to modify several columns, then Ctrl-click the requisite
   column titles. In this example, you could select the CostPrice and
   TotalDiscount columns.

4. Click the Data Type button in the Transform ribbon. A pop-up
   menu of potential data types will appear.

5. Select an appropriate data type. If you have selected the CostPrice
   column, then Whole Number is the type to choose.

After a few seconds, the data type will be applied. Changed Type will appear in the
Applied Steps list. The data types that you can apply are outlined in Table 10-1.

Table 10-1. *Data Types in Power BI Desktop*

Data Type	Description
Decimal Number	Converts the data to a decimal number.
Fixed Decimal Number	Converts the data to a decimal number with a fixed number of decimals.
Whole Number	Converts the data to a whole (integer) number.
Date/Time	Converts to a date and time data type.
Date	Converts to a date data type.
Time	Converts to a time data type.
Duration	Sets the data as being a duration. These are used for date and time calculations.
Text	Sets to a text data type.
True/False	Sets the data type to Boolean (true or false).
Binary	Defines the data as binary, and consequently, it is not directly visible.

Note   The Data Type button is also available in the Home ribbon. Equally, you can right-click a column header and select Change Type to select a different data type.

Inevitably, there will be times when you try to apply a data type that simply cannot be used with a certain column of data. Converting a text column (such as Make in this sample data table) into dates will simply not work. If you do this, then Power BI Desktop Query Editor will replace the column contents with Error. This is not definitive or dangerous, and all you have to do to return the data to its previous state is to delete the Changed Type step in the Applied Steps list by clicking the small cross at the left of the step name.

Sometimes, you could try and change a data type when the data type has already been changed. In this case, you will get an alert like the one shown in Figure 10-1.

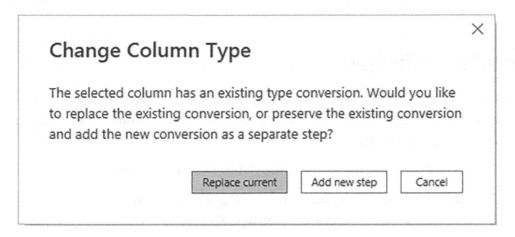

Figure 10-1. *The Change Column Type alert*

If this occurs, you can do one of two things:

- Let Power BI Desktop update the existing conversion step with the data type that you just selected.

- Add a new conversion step.

Your choice will depend on exactly what type of transformation you are applying to the underlying dataset.

It can help to alter data types at the same time for a *set* of columns where you think that this operation is necessary. There are a couple of good reasons for this approach:

- You can concentrate on getting data types right, and if you are working methodically, you are less likely to forget to set a data type.

- Applying data types for many columns (even if you are doing this in several operations, to single or multiple columns) will only add a single step to the Applied Steps list.

Note   Don't look for any data *formatting* options in Power Query; there aren't any. This is deliberate since this tool is designed to structure, load, and cleanse data, but not to present it. You carry out the formatting in the Power BI Desktop Data View, as you will see in Chapter 13.

Detecting Data Types

Applying the correct data type to dozens of columns can be more than a little time-consuming. Fortunately, Power BI Desktop now contains an option to apply data types automatically to a whole table:

1. In the Transform ribbon, click the Detect Data Type button.

2. Changed Type will appear in the Applied Steps list. Most of the columns will have the correct data type applied.

This technique does not always give perfect results, and there will be times when you want to override the choice of data type that Power BI Desktop has applied. Yet it is nonetheless a welcome addition to the data preparation toolset that can save you considerable time when preparing a dataset.

Data Type Indicators

It would be singularly unproductive to have to guess which column was set to which data type. So Power BI Desktop comes to your aid by indicating, visually, the corresponding data type for each column. If you look closely to the left of each individual column header, you will see a tiny icon. Each icon specifies the column's data type. The meaning of each icon is given in Table 10-2.

Table 10-2. *Data Type Icons in Power BI Desktop Query Editor*

Data Type Icon	Description
ABC 123	Any data type from among the possible data types
$1^2{}_3$	Whole number
1.2	Decimal number
$	Fixed decimal number
%	Percentage
$A^B{}_C$	Text
$\times_\vee$	True/false
Date/time icon	Date/time
Date icon	Date

(*continued*)

Table 10-2. (*continued*)

Data Type Icon	Description
⏱	Time
🌐	Date/time/time zone
⏱	Duration
📄	Binary

Switching Data Types

Another quick way to alter the data type for a column is to click the data type icon to the left of the column title and select the required data type from the context menu that you can see in Figure 10-2.

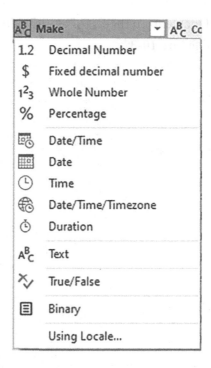

Figure 10-2. *The data type context menu*

You need to be aware, nonetheless, that while Power Query suggests all the available data types, this does not mean that a type conversion is actually possible.

Data Type Using Locale

When you are converting data types, you can also choose to use the current locale to specify date, time, and number formats. This means that users opening the Power BI Desktop file in another country will see date, time, and number formats adapted to the local formatting conventions. To do this:

1. Open Power Query (unless it is already open).

2. Click the data type icon to the left of the column title.

3. Select Using Locale from the pop-up menu. The Change Type with Locale dialog will appear.

4. Choose the new data type to apply from the list of available data types.

5. Select the required locale from the list of worldwide locales. The dialog will look like Figure 10-3.

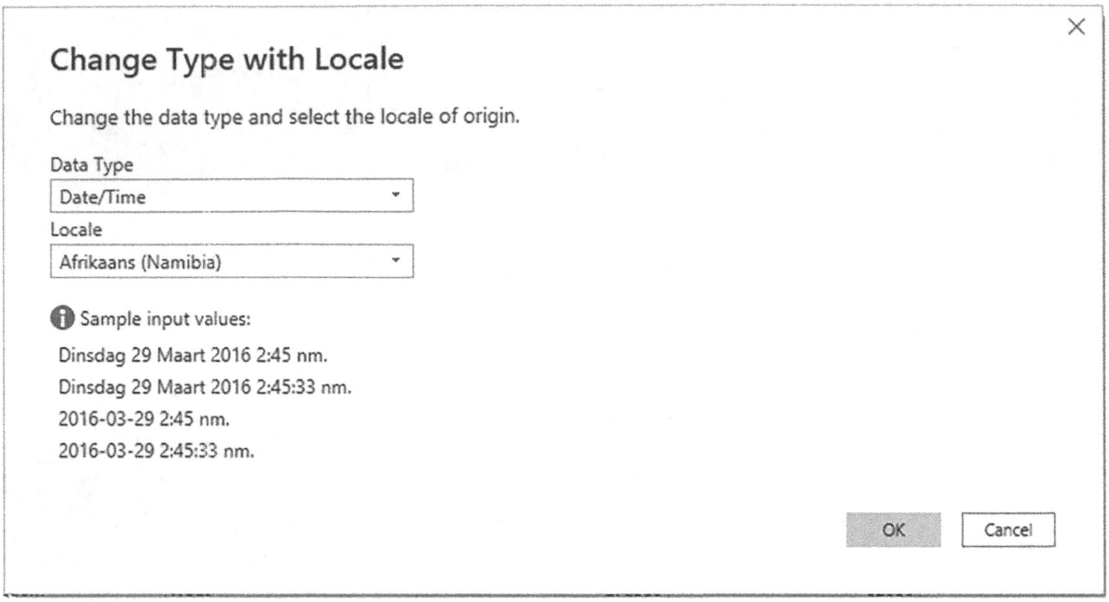

Change Type with Locale

Change the data type and select the locale of origin.

Data Type

Date/Time

Locale

Afrikaans (Namibia)

ⓘ Sample input values:

Dinsdag 29 Maart 2016 2:45 nm.
Dinsdag 29 Maart 2016 2:45:33 nm.
2016-03-29 2:45 nm.
2016-03-29 2:45:33 nm.

OK   Cancel

Figure 10-3. *The Change Type with Locale dialog*

6. Click OK.

The data type will be converted to the selected locale. The Applied Steps list will contain a step entitled Changed Type with Locale.

Replacing Values

Some data that you load will need certain values to be replaced by others in a kind of global search-and-replace operation—just as you would in a document. For instance, perhaps you need to standardize spellings where a make of car (to use the current sample dataset as an example) has been entered incorrectly. To carry out this particular data cleansing operation, do the following:

1. In Power Query, click the title of the column that contains the data that you want to replace. The column will become selected. In this example, I used the Model column from the file Sample1.pbix as an example.

2. In the Home ribbon, click the Replace Values button. The Replace Values dialog will appear.

3. In the Value To Find box, enter the text or number that you want to replace. I used Ghost in this example.

4. In the Replace With box, enter the text or number that you want to replace. I used Fantôme in this example, as shown in Figure 10-4.

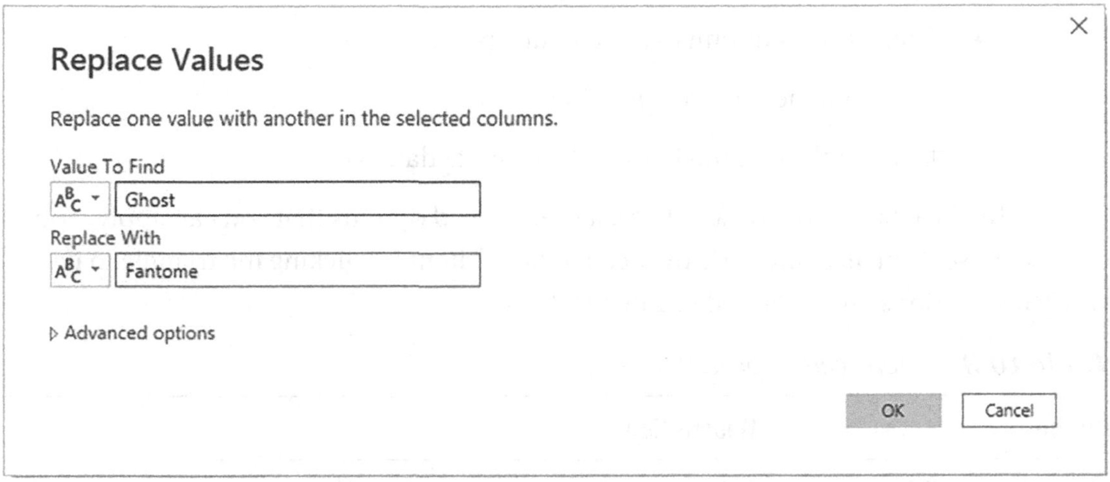

Figure 10-4. *The Replace Values dialog*

5. Click OK. The data is replaced in the entire column. Replaced Value is added to the Applied Steps list.

I only have a few comments about this technique:

- The Replace Values process searches for every occurrence of the text that you are looking for in each record of the selected columns. It does not look for the entire contents of the cell unless you specifically request this by checking the Match Entire Cell Contents check box in the advanced options.

- If you click a cell containing the contents that you want to replace (rather than the column title, as we just did), before starting the process, Power BI Desktop automatically places the cell contents in the Replace Values dialog as the value to find.

- You can only replace text in columns that contain text elements. This does not work with columns that are set as a numeric or date data type. Indeed, you will see a yellow alert triangle in the Replace Values dialog if you enter values that do not match the data type of the selected column(s).

- If you really have to replace parts of a date or figures in a numeric column with other dates or numbers, then you can

  - Convert the column to a text data type

  - Carry out the replace operation

  - Convert the column back to the original data type

The Replace Values dialog also has a few advanced options that you can apply. You can see these if you expand the "Advanced options" item by clicking the triangle to its left. These options are explained in Table 10-3.

Table 10-3. *Advanced Replace Options*

Option	Description
Match Entire Cell Contents	Only replaces the search value if it makes up the entire contents of the cell
Replace Using Special Characters	Replaces the search value with a nonprinting character
Tab	Replaces the search value with a tab character
Carriage Return	Replaces the search value with a carriage return character
Line Feed	Replaces the search value with a line feed character
Carriage Return and Line Feed	Replaces the search value with a carriage return and line feed

Note   Replacing words that are subsets of other words is dangerous. When replacing any data, make sure that you don't damage elements other than the one you intend to change.

As a final and purely spurious comment, I must add that I would never suggest rebranding a Rolls Royce, as it would be close to automotive sacrilege.

Transforming Column Contents

Power BI Desktop has a powerful toolbox of automated data transformations that allow you to standardize the contents of a column in several ways. These include

- Setting the capitalization of text columns

- Rounding numeric data or applying math functions

- Extracting date elements such as the year, month, or day (among others) from a date column

Power BI Desktop is very strict about applying transformations to appropriate types of data. This is because transforms are totally dependent on the data type of the selected column. This is yet another confirmation that applying the requisite data type is an operation that should be carried out early in any data transformation process—and certainly *before* transforming the column contents. Remember, you will only be able to select a numeric transformation if the column is a numeric data type, and you will only be able to select a date transformation if the column is a date data type. Equally, the text-based transformations can only be applied to columns that are of the text data type.

Text Transformation

Let's look at a simple transformation operation in action. As an example, I will get Power BI Desktop to convert the Make column into uppercase characters.

1. Still using the file Example1.pbix (and having switched to Query Editor), click anywhere in the column whose contents you wish to transform (Make in this case).

2. In the Transform ribbon, click the Format button. A pop-up menu will appear.

3. Select UPPERCASE, as shown in Figure 10-5.

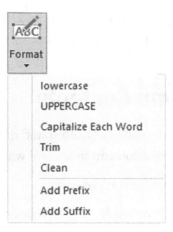

Figure 10-5. *The Format menu*

The contents of the entire column will be converted to uppercase. Uppercased Text will be added to the Applied Steps list.

As you can see from the menu for the Format button, you have five possible options when formatting (or transforming) text. These options are explained in Table 10-4.

Table 10-4. *Text Transformations*

Transformation	Description	Applied Steps Definition
Lowercase	Converts all the text to lowercase	Lowercased Text
Uppercase	Converts all the text to uppercase	Uppercased Text
Capitalize Each Word	Converts the first letter of each word to a capital	Capitalized Each Word
Trim	Removes all spaces before and after the text	Trimmed Text
Clean	Removes any nonprintable characters	Cleaned Text
Add Prefix	Adds text at the start of the column contents	Added Prefix
Add Suffix	Adds text at the end of the column contents	Added Suffix

Note   I realize that Power Query calls text transformations formatting. Nonetheless, these options are part of the overall data transformation options.

Adding a Prefix or a Suffix

You can also add a prefix or a suffix to all the data in a column. This is as easy as the following:

1. In Power Query, click inside the column where you want to add a prefix.

2. In the Transform ribbon, select Format ➤ Add Prefix. The Prefix dialog will be displayed, as you can see in Figure 10-6.

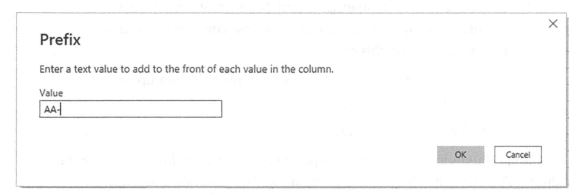

Figure 10-6. *Adding a prefix to a text*

3. Enter the prefix to add in the Value field.

4. Click OK.

The prefix that you designated will be placed at the start of every record in the dataset for the selected field.

Note   If you add a prefix or a suffix to a numeric or date/time column, then the column data type will automatically be converted to text.

Removing Leading and Trailing Spaces

There will inevitably be occasions when you inherit data that has extra spaces before, after, or before *and* after the data itself. This can be insidious, as it can cause

- Data duplication, because a value with a trailing space is *not* considered identical to the same text without the spaces that follow

- Sort issues, because a leading space causes an element to appear at the *top* of a sorted list

- Grouping errors, because elements with spaces are not part of the same group as elements without spaces

Fortunately, Power Query has a ruthlessly efficient solution to this problem.

1. Still using the file Example1.pbix (and having switched to Query Editor), click anywhere in the column whose contents you wish to transform (Make in this case).

2. In the Transform ribbon, click the Format button. A pop-up menu will appear.

3. Select Trim from the menu.

All superfluous leading and trailing spaces will be removed from the data in the column. This can help when sorting, grouping, and deduplicating records.

Removing Nonprinting Characters

Some source data can contain somewhat insidious elements called nonprinting characters. These can, even if they are nearly always invisible to humans, cause problems when you print reports and dashboards.

If you suspect that your source data contains nonprinting characters, you can remove them simply like this:

1. In Power Query, click inside the column (or select the columns) that you know to contain (or that you suspect contain) nonprinting characters.

2. Click Format ➤ Clean.

Power BI Desktop will add Cleaned Text to the list of Applied Steps.

Number Transformations

Just as you can transform the contents of text-based columns, you can also apply transformations to numeric values. As an example, suppose that you want to round up all the figures in a column to the nearest whole number.

1. In Power Query, click anywhere in the column whose contents you wish to transform (TotalDiscount in this case).

2. In the Transform ribbon, click the Rounding button. A pop-up menu will appear showing all the available options. You can see this in Figure 10-7.

Figure 10-7. *Rounding options*

3. Select Round Up.

The values in the entire column will be rounded up to the nearest whole number. Rounded Up will be added to the Applied Steps list.

The other possible numeric transformations that are available are described in Table 10-5. Because these numeric transformations use several buttons in the Transform ribbon, I have indicated which button to use to get the desired result.

Table 10-5. *Number Transformations*

Transformation	Description	Applied Steps Definition
Rounding ➤ Round Up	Rounds each number up to the specified number of decimal places.	Rounded Up
Rounding ➤ Round Down	Rounds each number down.	Rounded Down
Round…	Rounds each number to the number of decimals that you specify. If you specify a negative number, you round to tens for -1, hundreds for -2, etc.	Rounded Off
Scientific ➤ Absolute Value	Makes the number absolute (positive).	Calculated Absolute Value
Scientific ➤ Power ➤ Square	Returns the square of the number in each cell.	Calculated Square
Scientific ➤ Power ➤ Cube	Returns the cube of the number in each cell.	Calculated Cube
Scientific ➤ Power ➤ Power	Raises each number to the power that you specify.	Calculated Power
Scientific ➤ Square Root	Returns the square root of the number in each cell.	Square Root
Scientific ➤ Exponent	Returns the exponent of the number in each cell.	Calculated Exponent
Scientific ➤ Logarithm ➤ Base 10	Returns the base 10 logarithm of the number in each cell.	Calculated Base 10 Logarithm
Scientific ➤ Logarithm ➤ Natural	Returns the natural logarithm of the number in each cell.	Calculated Natural Logarithm
Scientific ➤ Factorial	Gives the factorial of numbers in the column.	Calculated Factorial
Trigonometry ➤ Sine	Gives the sine of the numbers in the column.	Calculated Sine

(continued)

Table 10-5. (*continued*)

Transformation	Description	Applied Steps Definition
Trigonometry ➤ Cosine	Gives the cosine of the numbers in the column.	Calculated Cosine
Trigonometry ➤ Tangent	Gives the tangent of the numbers in the column.	Calculated Tangent
Trigonometry ➤ ArcSine	Gives the arcsine of the numbers in the column.	Calculated Arcsine
Trigonometry ➤ ArcCosine	Gives the arccosine of the numbers in the column.	Calculated Arccosine
Trigonometry ➤ ArcTangent	Gives the arctangent of the numbers in the column.	Calculated Arctangent

Note   Power Query will not even let you try to apply numeric transformation to texts or dates. The relevant buttons remain grayed out if you click inside a column of letters or dates.

Calculating Numbers

Power Query can also apply simple arithmetic to the figures in a column. Suppose, for instance, that you want to multiply all the sale prices by 110% as part of your forecasts. This is how you can do just that:

1. Still using the file Example1.pbix (and having switched to Query Editor), click inside any column of numbers. In this example, I used the column SalePrice.

2. Click the Standard button in the Transform ribbon. The menu will appear as you can see in Figure 10-8.

Figure 10-8. *Applying a calculation to a column*

3. Click Multiply. The Multiply dialog will appear.

4. Enter **1.1** in the Value box. The dialog will look like the one shown in Figure 10-9.

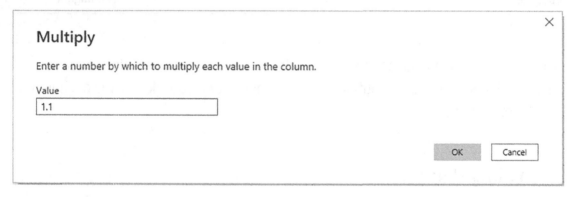

Figure 10-9. *Applying a calculation to a column*

5. Click OK.

All the numbers in the selected column will be multiplied by 1.1. In other words, they are now 110% of the original value. Table 10-6 describes the possible math operations that you can carry out in Power Query.

Table 10-6. *Applying Basic Calculations*

Transformation	Description	Applied Steps Definition
Add	Adds a selected value to the numbers in a column	Added to Column
Multiply	Multiplies the numbers in a column by a selected value	Multiplied Column
Subtract	Subtracts a selected value from the numbers in a column	Subtracted from Column
Divide	Divides the numbers in a column by a selected value	Divided Column
Integer-Divide	Divides the numbers in a column by a selected value and removes any remainder	Integer-Divided Column
Modulo	Divides the numbers in a column by a selected value and leaves only the remainder	Calculated Modulo
Percentage	Applies the selected percentage to the column	Calculated Percentage
Percent Of	Expresses the value in the column as a percent of the value that you enter	Calculated Percent Of

Note   You can also carry out many types of calculations in Power BI Desktop Data View and avoid carrying out calculations in Power Query. Indeed, many Power BI Desktop purists seem to prefer that anything resembling a calculation should take place inside the data model rather than at the query stage. As ever, I will let you decide which approach you prefer. Be aware that some heavy transforms can slow the reports down if calculated at runtime, whereas others can only be effective as part of a well-thought-out calculation process.

Finally, it is important to remember that you are altering the data when you carry out this kind of operation. In the real world, you might be safer duplicating a column before profoundly altering the data it contains. This allows you to keep the initial data available, albeit at the cost of increasing both the load time and the size of the Power BI Desktop file.

Date Transformations

Transforming dates follows similar principles to transforming text and numbers. As an example, here is how to isolate the month from a date:

1. Still using the file Example1.pbix (and having switched to Query Editor), click inside the InvoiceDate column.

2. In the Transform ribbon, click the Date button. The menu will appear.

3. Click Year. The submenu will appear.

4. Select Year. The year part of the date will replace all the dates in the InvoiceDate column.

The other possible date transformations that are possible are given in Table 10-7.

Table 10-7. *Date Transformations*

Transformation	Description	Applied Steps Definition
Age	Calculates the date and time difference (in days and hours) between the original date and the current local time	Calculated Age
Date Only	Converts the data to a date without the time element	Extracted Date
Year ➤ Year	Extracts the year from the date	Extracted Year
Year ➤ Start of Year	Returns the first day of the year for the date	Extracted Start of Year
Year ➤ End of Year	Returns the last day of the year for the date	Extracted End of Year
Month ➤ Month	Extracts the number of the month from the date	Extracted Month
Month ➤ Start of Month	Returns the first day of the month for the date	Extracted Start of Month
Month ➤ End of Month	Returns the last day of the month for the date	Extracted End of Month

(continued)

Table 10-7. (*continued*)

Transformation	Description	Applied Steps Definition
Month ➤ Days in Month	Returns the number of days in the month for the date	Extracted Days in Month
Month ➤ Name of Month	Returns the name of the month for the date	Extracted Month Name
Day ➤ Day	Extracts the day from the date	Extracted Day
Day ➤ Day of Week	Returns the weekday as a number (Monday is 0, Tuesday is 1, etc.)	Extracted Day of Week
Day ➤ Day of Year	Calculates the number of days since the start of the year for the date	Extracted Day of Year
Day ➤ Start of Day	Transforms the value to the start of the day for a date and time	Extracted Start of Day
Day ➤ End of Day	Transforms the value to the end of the day for a date and time	Extracted End of Day
Day ➤ Name of Day	Returns the weekday as a day of week	Extracted Day Name
Quarter ➤ Quarter of Year	Returns the calendar quarter of the year for the date	Quarter of Year
Quarter ➤ Start of Quarter	Returns the first date of the calendar quarter of the year for the date	Extracted Start of Quarter
Quarter ➤ End of Quarter	Returns the last date of the calendar quarter of the year for the date	Extracted End of Quarter
Week ➤ Week of Year	Calculates the number of weeks since the start of the year for the date	Extracted Week of Year
Week ➤ Week of Month	Calculates the number of weeks since the start of the month for the date	Extracted Week of Month
Week ➤ Start of Week	Returns the date for the first day of the week (Monday) for the date	Calculated Start of Week
Week ➤ End of Week	Returns the date for the last day of the week (Sunday) for the date	Calculated End of Week

315

Time Transformations

You can also transform date/time or time values into their component parts using Power Query. This is extremely similar to how you apply date transformations, but in the interest of completeness, the following explains how to do this:

1. Click inside the InvoiceDate column.

2. In the Transform ribbon, click the Time button. The menu will appear.

3. Click Hour. The hour part of the time will replace all the values in the InvoiceDate column.

Note   Time transformations can only be applied to columns of the date/time, time, or time zone data types.

The range of time transformations is given in Table 10-8.

Table 10-8. *Time Transformations*

Transformation	Description	Applied Steps Definition
Time Only	Isolates the time part of a date and time	Extracted Time
Local Time	Converts the date/time to local time from date/time and time zone values	Calculated Local Time
Parse	Extracts the date and/or date/time elements from a text	Parsed Date (or Parsed Time)
Hour ➤ Hour	Isolates the hour from a date/time or datetime value	Extracted Hour
Hour ➤ Start of Hour	Returns the start of the hour from a date/time or time value	Calculated Start of Hour
Hour ➤ End of Hour	Returns the end of the hour from a date/time or time value	Calculated End of Hour

(continued)

Table 10-8. (*continued*)

Transformation	Description	Applied Steps Definition
Minute	Isolates the minute from a date/time or time value	Extracted Minute
Second	Isolates the second from a date/time or time value	Extracted Second
Earliest	Returns the earliest time from a date/time or time column	Calculated Earliest
Latest	Returns the latest time from a date/time or time column	Calculated Latest

Note   In the real world, you could well want to leave a source column intact and apply number or date transformations to a copy of the column. To do this, simply apply the same transformation technique; only use the buttons in the Add Column ribbon instead of those in the Transform ribbon.

One final time (or rather datetime) modification is that if you select both a date column and a time column (such as InvoiceDateOnly and InvoiceTime), you will see the option Combine Date and Time in the Time button pop-up menu. This combines a date column and a time column into a single datetime column.

Duration

If you have values in a column that can be interpreted as a duration (in days, hours, minutes, and seconds), then Power BI Desktop Query can extract the component parts of the duration as a data transformation. For this to work, however, the column *must* be set to the duration data type. This means that the contents of the column have to be interpreted as a duration by Power BI Desktop. Any values that are incompatible with this data type will be set to error values.

A duration string could look like the following: **55.11:22:33.5000000**.

In this case:

- **55** is the number of days.

- **11** is the number of hours.

- **22** is the number of minutes.

- **33** is the number of seconds.

- **50000000** is the number of milliseconds.

Note that a decimal separates the number of days from the hours, colons separate hours from minutes and minutes from seconds, and another decimal separates seconds from milliseconds.

If you have duration data, you can extract its component parts like this:

1. Still using the file Example1.pbix (and having switched to Query Editor), click inside the column InvoiceDate.

2. In the Transform ribbon, click the Duration button. The menu will appear.

3. Click Hour. The hour part of the time will replace all the values in the InvoiceDate column.

The range of duration transformations is given in Table 10-9.

Table 10-9. *Duration Transformations*

Transformation	Description	Applied Steps Definition
Days	Isolates the day element from a duration value	Extracted Days
Hours	Isolates the hour element from a duration value	Extracted Hours
Minutes	Isolates the minute element from a duration value	Extracted Minutes
Seconds	Isolates the second element from a duration value	Extracted Seconds
Total Years	Displays the duration value as the number of years and a fraction representing days, minutes, and seconds	Calculated Total Years
Total Days	Displays the duration value as the number of days and a fraction representing hours, minutes, and seconds	Calculated Total Days
Total Hours	Displays the duration value as the number of hours and a fraction representing minutes and seconds	Calculated Total Hours

(continued)

Table 10-9. (*continued*)

Transformation	Description	Applied Steps Definition
Total Minutes	Displays the duration value as the number of minutes and a fraction representing seconds	Calculated Total Minutes
Total Seconds	Displays the duration value as the number of seconds and a fraction representing milliseconds	Calculated Total Seconds
Multiply	Multiplies the duration (and all its component parts) by a value that you enter	Multiplied Column
Divide	Divides the duration (and all its component parts) by a value that you enter	Divided Column
Statistics ➤ Sum	Returns the total for all the duration elements in the column	Calculated Sum
Statistics ➤ Minimum	Returns the minimum value of all the duration elements in the column	Calculated Minimum
Statistics ➤ Maximum	Returns the maximum value of all the duration elements in the column	Calculated Maximum
Statistics ➤ Median	Returns the median value for all the duration elements in the column	Calculated Median
Statistics ➤ Average	Returns the average for all the duration elements in the column	Calculated Average

Note   If you multiply or divide a duration, Power Query displays a dialog so that you can enter the value to multiply or divide the duration by.

Conclusion

In this chapter, you learned to apply many of the core data cleansing techniques that Power Query can use to render data more usable. This included changing data types, changing case, replacing values, and removing extraneous white space. You also saw how to remove hidden characters that can prevent accurate data display.

Then you learned to apply mathematical, statistical, and trigonometric functions to numbers as they are imported. Other techniques covered extracting date, time, and duration elements from date/time and duration columns.

It is now time to see how you can carry out a series of core data transformations, including extracting part of the data in a column, merging the data from two or more columns, and separating out the data from one column into other columns—among many more useful data transformation processes. All of this will be the subject of the next chapter.

CHAPTER 11

Data Transformation

Not all data is perfect. You may need to make a few tweaks or even spend hours reworking the entire dataset to make the original source data usable in your dashboards. Generically, this is known as *data transformation*. Transforming data is one area where Power Query excels, and this chapter will introduce you to a range of techniques that you may one day need to apply to adjust (or pummel) your data sources so that they form a solid basis for analytics.

So what does data transformation entail? The term covers a range of approaches and functions that cover

- Filling data down or up over empty cells to ensure that records are complete

- Extracting part of the data in a column into a new column

- Separating all the data in a column so that each data element appears in a separate column

- Merging columns into a new column

- Adding custom columns that possibly contain calculations or extract part of a column's data into a new column or even concatenate columns

- Adding "index" columns to ensure uniqueness or memorize a sort order

These are the main techniques that you will learn in this chapter. You will notice that, as most of these operations are concerned with *changing* data, many of them are accessed using the Transform ribbon. Also, it is worth noting that cleansing data does not only consist of reducing it. Sometimes, you may have to *extend* the data to make it usable. This normally means adding further columns to a data table.

© Adam Aspin 2022
A. Aspin, *Pro Data Mashup for Power BI*, https://doi.org/10.1007/978-1-4842-8578-7_11

The sample data that you will need to follow the exercises in this chapter is in the folder C:\PowerBiDesktopSamples. Please note that many of the examples in this chapter presume that you have already opened Power Query.

Filling Down Empty Cells

Imagine a data source where the data has come into Power BI Desktop from a matrix-style structure. The result is that some columns only contain a single example of an element and then a series of empty cells until the next element in the list. If this is difficult to imagine, then take a look at the sample file CarMakeAndModelMatrix.xlsx shown in Figure 11-1.

Make	Marque	Sales
Aston Martin	DB4	391000
	DB7	500740
	DB9	915070
	DBS	230000
	Rapide	225000
	Vanquish	746500
	Vantage	320850
	Zagato	178500
Bentley	Arnage	44000
	Azure	239250
	Continental	991250
	Turbo R	347500
Jaguar	XJ12	303500
	XJ6	602000
	XK	1092250
MGB	GT	315000
Rolls Royce	Camargue	810300
	Phantom	178500
	Silver Ghost	649500
	Silver Seraph	288500
	Silver Shadow	308500
	Wraith	178500
Triumph	TR4	140500
	TR5	98250
	TR7	47750
TVR	Cerbera	89250
	Tuscan	112250

Figure 11-1. *A matrix data table in Excel*

All these blank cells are a problem since we need a full data table without empty cells for Power BI to create accurate visuals. Or rather, they would be a problem if Power BI Desktop did not have a really cool way of overcoming this particular difficulty. Do the following to solve this challenge:

1. Open a new Power BI Desktop file.

2. In the splash screen, click Get Data.

3. In the Get Data dialog, select Excel. Then click Connect and navigate to `C:\PowerBiDesktopSamples\CarMakeAndModelMatrix.xlsx`.

4. Click Open, select Sheet 1, and click Transform Data. This will take you directly to Power Query.

5. Select the column that contains the empty cells (the Make column in this example).

6. In the Transform ribbon, click Fill. The menu will appear as shown in Figure 11-2.

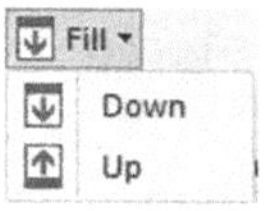

Figure 11-2. *The Fill menu*

7. Select Down. The blank cells will be replaced by the value in the first nonempty cell above. Filled Down will be added to the Applied Steps list.

The table will now look like Figure 11-3.

⊞▾	A^B_C Make	▾	A^B_C Marque	▾	1$^2_3$ Sales	▾
1	Aston Martin		DB4		391000	
2	Aston Martin		DB7		500740	
3	Aston Martin		DB9		915070	
4	Aston Martin		DBS		230000	
5	Aston Martin		Rapide		225000	
6	Aston Martin		Vanquish		746500	
7	Aston Martin		Vantage		320850	
8	Aston Martin		Zagato		178500	
9	Bentley		Arnage		44000	
10	Bentley		Azure		239250	
11	Bentley		Continental		991250	
12	Bentley		Turbo R		347500	
13	Jaguar		XJ12		303500	
14	Jaguar		XJ6		602000	
15	Jaguar		XK		1092250	
16	MGB		GT		315000	
17	Rolls Royce		Camargue		810300	
18	Rolls Royce		Phantom		178500	
19	Rolls Royce		Silver Ghost		649500	
20	Rolls Royce		Silver Seraph		288500	
21	Rolls Royce		Silver Shadow		308500	
22	Rolls Royce		Wraith		178500	
23	Triumph		TR4		140500	
24	Triumph		TR5		98250	
25	Triumph		TR7		47750	
26	TVR		Cerbera		89250	
27	TVR		Tuscan		112250	

Figure 11-3. *A data table with empty cells replaced by the correct data*

Note   This technique is built to handle a fairly specific problem and only really works if the imported data is grouped by the column containing the missing elements.

Although rare, you can also use this technique to fill empty cells with the value from below. If you need to do this, just select Fill ➤ Up from the Transform ribbon. In either case, you need to be aware that the technique is applied to the entire column.

Extracting Part of a Column's Contents

There could well be times when the contents of a source column contain more data than you actually need. In cases like this, Power BI Desktop can help you by extracting only part of a column. This technique works like this:

1. Load the C:\PowerBiDesktopSamples\Example1.pbix sample file and click Transform Data to open Power Query.

2. Click inside the InvoiceNumber column.

3. In the Transform ribbon, click Extract ➤ Text Before Delimiter. The Text Before Delimiter dialog will be displayed.

4. Enter a hyphen (or a minus sign) in the Delimiter field. The dialog will look like Figure 11-4.

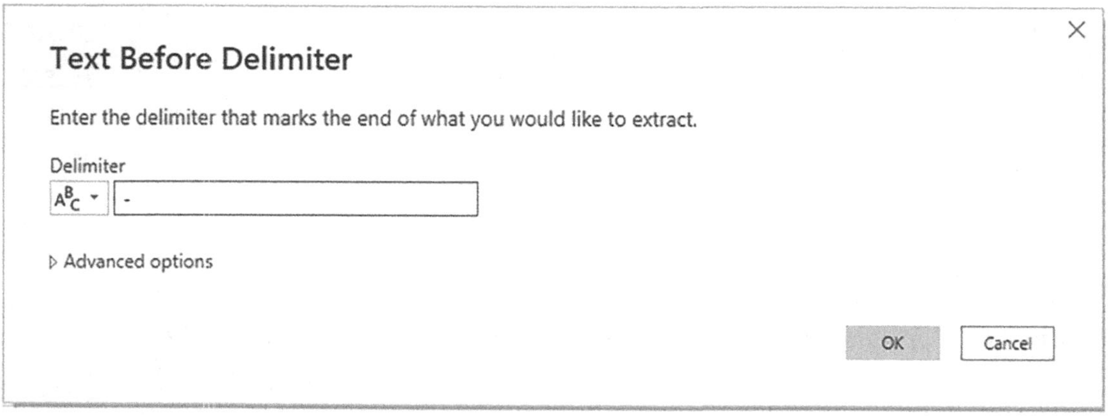

Figure 11-4. The Text Before Delimiter dialog

5. Click OK. The contents of the field will be replaced by the characters before the hyphen. A step named Extracted Text Before Delimiter will be added to the Applied Steps list.

The Extract function allows you to choose from a variety of ways in which you can extract a subset of data from a column. The currently available options are explained in Table 11-1.

Table 11-1. *Extract Transformations*

Transformation	Description	Applied Steps Definition
Length	Displays the length in characters of the contents of the field	Calculated Text Length
First Characters	Displays a specified number of characters from the left of the field	Extracted First Characters
Last Characters	Displays a specified number of characters from the right of the field	Extracted Last Characters
Range	Displays a specified number of characters between a specified start and end position (in characters, from the left of the field)	Extracted Text Range
Text Before Delimiter	Displays all the text occurring before a specified character	Extracted Text Before Delimiter
Text After Delimiter	Displays all the text occurring after a specified character	Extracted Text After Delimiter
Text Between Delimiters	Displays all the text occurring between two specified characters	Extracted Text Between Delimiters

Advanced Extract Options

Three of the Extract options (Text Before Delimiter, Text After Delimiter, and Text Between Delimiters) let you apply some advanced options that allow you to push the envelope even further when extracting data from a column. These techniques are explained in the following two sections.

Text Before and After Delimiter

If you are extracting part of the contents of a column and you are using a delimiter to isolate the text you want to keep, then you have a couple of additional options available.

You can access these options from the dialog that you saw in Figure 11-4 by clicking Advanced options. The dialog will then look like the one shown in Figure 11-5.

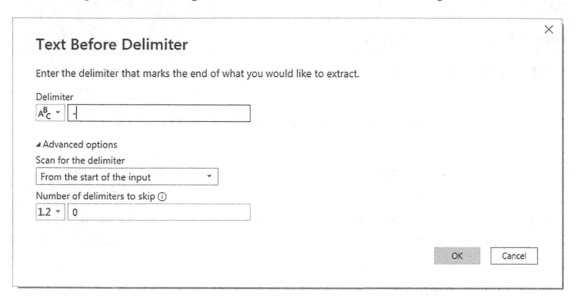

Figure 11-5. *The Advanced options of the Text Before and Text After Delimiter dialogs*

The two options that you now have are

- *Scan for the delimiter*: This option lets you choose between working forward from the start of the contents of the column and working backward from the end of the contents of the column to locate the delimiter you are searching for.

- *Number of delimiters to skip*: Here, you can specify that it is the *n+1*th occurrence of a delimiter that interests you.

Text Between Delimiters

The Advanced options of the Text Between Delimiters dialog essentially lets you apply the same options that you saw previously, only for both the initial delimiter and the final delimiter. In Figure 11-6, you can see this in the Text Between Delimiters dialog.

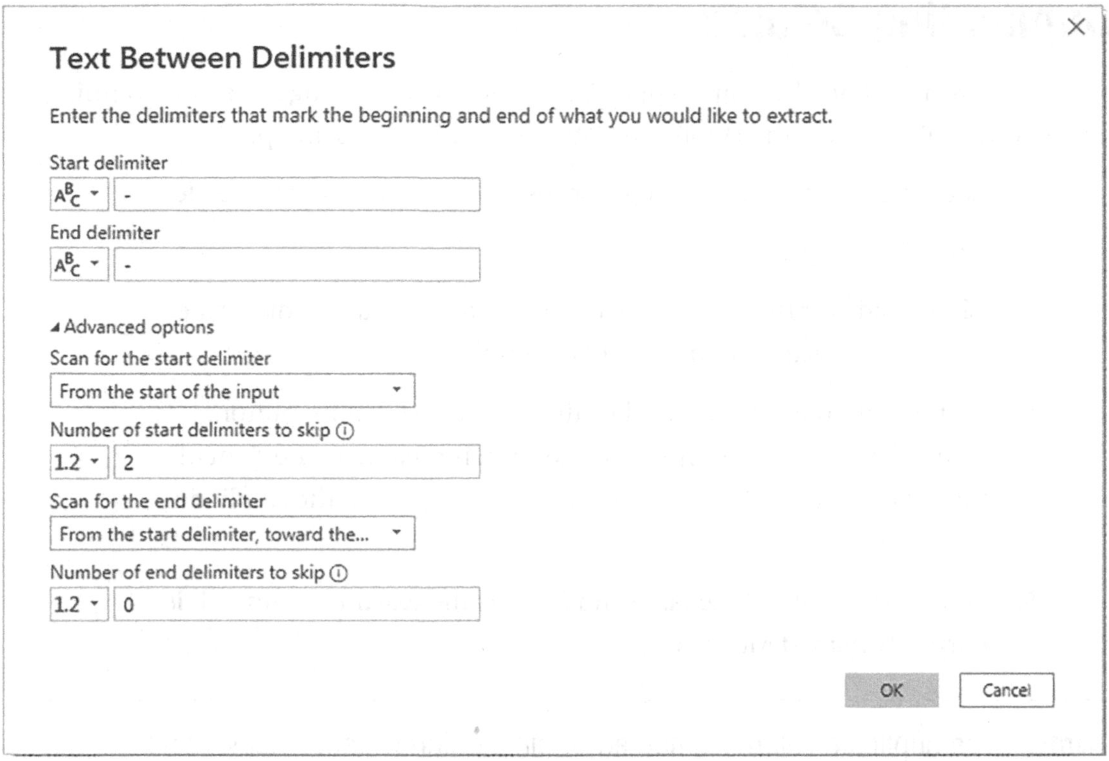

Text Between Delimiters

Enter the delimiters that mark the beginning and end of what you would like to extract.

Start delimiter

A^B_C ▾ | -

End delimiter

A^B_C ▾ | -

◢ Advanced options

Scan for the start delimiter

From the start of the input   ▾

Number of start delimiters to skip ⓘ

1.2 ▾ | 2

Scan for the end delimiter

From the start delimiter, toward the...   ▾

Number of end delimiters to skip ⓘ

1.2 ▾ | 0

OK   Cancel

Figure 11-6. *The Advanced options of the Text Between Delimiters dialog*

Note   The Extract button can be found in both the Transform and New Column ribbons. If you carry out this operation from the Transform ribbon, then the contents of the existing column will be replaced. If you use the button in the Add Column ribbon, then a new column containing the extracted text will be added at the right of any existing columns.

The start and end delimiters, the number of start delimiters to skip, and the number of end delimiters to skip can be values (as was the case in this example) or parameters (which you will learn about in Chapter 13).

Duplicating Columns

Sometimes, you just need a simple copy of a column, with nothing added and nothing taken away. This is where the Duplicate Column button comes into play.

1. Load the `C:\PowerBiDesktopSamples\Example1.pbix` sample file.

2. Open Power Query.

3. Click inside (or the title of) the column that you want to duplicate. I will use the Make column in this example.

4. In the Add Column ribbon, click the Duplicate Column button. After a few seconds, a copy of the column is created at the right of the existing table. Duplicated Column will appear in the Applied Steps list.

5. Scroll to the right of the table and rename the existing column; it is currently named Make - Copy.

Note   The duplicate column is named Original Column Name-Copy. I find that it helps to rename copies of columns sooner rather than later in a data transformation process.

Splitting Columns

Sometimes, a source column contains data that you really need to break up into smaller pieces across two or more columns. The following are classic cases where this happens:

- A column contains a list of elements, separated by a specific character (known as a *delimiter*).

- A column contains a list of elements, but the elements can be divided at specific positions in the column.

- A column contains a concatenated text that needs to be split into its composite elements (a bank account number and a Social Security number are examples of this).

The following short sections explain how to handle such eventualities.

Splitting Column by a Delimiter

Here is another requirement that you may encounter occasionally. The data that has been imported has a column that needs to be further split into multiple columns. Imagine a source file where a column holds a comma-separated list of elements. Once you have imported the file, you then need to further separate the contents of this column into separate columns.

Here is what you can do to split the data from one column over several columns:

1. Load the C:\PowerBiDesktopSamples\DataToParse.xlsx sample file and switch to Power Query.

2. In the Transform ribbon, click Use First Row as Headers. This promotes the initial record to be the column name as is sometimes necessary.

3. Click inside the ClientList column. You can see that this column contains several data elements, each separated by a semicolon.

4. In the Transform ribbon, click Split Column. You can see the Split Column menu options in Figure 11-7.

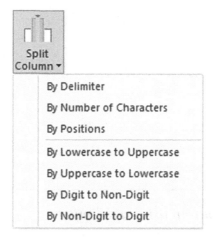

Figure 11-7. *The Split Column menu options*

5. Select By Delimiter. The Split Column by Delimiter dialog appears.

6. Select Semicolon from the list of available options in the "Select or enter delimiter" popup (although Power Query could well have detected this already).

7. Click "Each occurrence of the delimiter" as the location to split the text column. The dialog (with the Advanced options expanded for information only) should look like Figure 11-8.

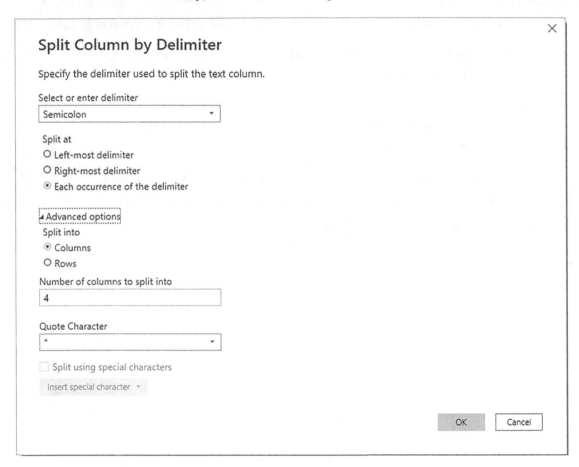

Figure 11-8. *Splitting a column using a delimiter*

8. Click OK. Split Column by Delimiter will appear in the Applied Steps list.

The initial column is replaced, and all the new columns are named ClientList.1, ClientList.2, and so forth. As many additional columns as there are delimiters are created; each is named (*Column.n*) and is sequentially numbered. The result of this operation looks like Figure 11-9.

A^B_C ClientList.1	A^B_C ClientList.2	A^B_C ClientList.3	A^B_C ClientList.4
Aldo Motors	Uttoxeter	Staffs	ST17 99RZ
Honest John	London		NSW1 1A
Bright Orange	Birmingham	NULL	B1 50AZ
Cut'n'Shut	Manchester	NULL	M1 5AZ
Wheels'R'Us	London	NULL	SE1 4YY
Les Arnaqueurs	Paris	NULL	75010
Crippen & Co	Glasgow	NULL	G1 8GH
Rocky Riding	New York	New York	NULL
Voitures Diplomatiques S.A.	Geneva	NULL	NULL
Karz	Stuttgart	NULL	NULL
Costa Del Speed	Madrid	NULL	NULL
Olde Englande	Shrewsbury	NULL	SY10 9AX
Impressive Wheels	Liverpool	NULL	L5 9ZZ
Smooth Riders	Telford	NULL	TF6 9RR
Luxury Rentals	Gloucester	NULL	GL7 9AS
Premium Motor Vehicles	Newcastle upon Tyne	NULL	NE3 3SS
Chateau Moi	Lyon	NULL	69001
Vive la Vitesse!	Marseille	NULL	13002
Carosse Des Papes	Avignon	NULL	84000
Three Country Cars	Basle	NULL	NULL
Jungfrau	Zurich	NULL	NULL
Ambassador Cars	Bellevue	Washington	NULL
Embassy Motors	Denver	Colorado	NULL
Style 'N Ride	Chevy Chase	Maryland	NULL
BritWheels	Portland	Oregon	NULL
Sporty Types Corp	San Francisco	California	NULL
Tweedy Wheels	Cambridge	Massahussets	NULL
Union Jack Sports Cars	Louisville	Kentucky	NULL
Buckingham Palace Car Services	Mason	Ohio	NULL
British Luxury Automobile Corp	Franklin	Tennesee	NULL
Classy Car Sales	Pittsburgh	Pennsylvania	NULL

Figure 11-9. *The results of splitting a column*

This particular process has several options, and their consequences can be fairly far-reaching as far as the data is concerned. Table 11-2 contains a description of the available options.

Table 11-2. *Delimiter Split Options*

Option	Description
Colon	Uses the colon (:) as the delimiter
Comma	Uses the comma (,) as the delimiter
Equals Sign	Uses the equals sign (=) as the delimiter
Semi-Colon	Uses the semicolon (;) as the delimiter
Space	Uses the space ( ) as the delimiter
Tab	Uses the tab character as the delimiter
Custom	Lets you enter a custom delimiter
At the Left-Most Delimiter	Splits the column once only at the first occurrence of the delimiter
At the Right-Most Delimiter	Splits the column once only at the last occurrence of the delimiter
At Each Occurrence of the Delimiter	Splits the column into as many columns as there are delimiters

Advanced Options for Delimiter Split

There are a small number of advanced options that are available when splitting text by delimiters. These are displayed when you click the Advanced options element in the Split Column by Delimiter dialog and are explained in Table 11-3.

Table 11-3. *Delimiter Split Advanced Options*

Advanced Options ➤ Number of Columns to Split Into	Allows you to set a maximum number of columns into which the data is split in chunks of the given number of characters. Any extra columns are placed in the rightmost column.
Advanced Options ➤ Quote Character	Separators inside a text that is contained in double quotes are not used to split the text into columns. Setting this option to "none" will split elements inside quotes.
Split using special characters	Enables the Insert Special Character button. You can then click this button and select the special character to split data on. The choice is between Tab, Carriage Return, Line-Feed, Carriage Return, and Line-Feed or nonbreaking space.

Splitting Columns by Number of Characters

Another variant on this theme is when text in each column is a fixed number of characters and needs to be broken down into constituent parts at specific intervals. Suppose, for instance, that you have a field where each group of (a certain number of) characters has a specific meaning, and you want to break it into multiple columns. Alternatively, suppose you want to extract the leftmost or rightmost n characters and leave the rest. A bank account and a Social Security number are examples of this. This is where splitting a column by the number of characters can come in useful. As the principle is very similar to the process that we just saw, I will not repeat the whole thing again. All you have to do is choose the "By number of characters" menu option at step 5 in the previous exercise. Options for this type of operation are given in Table 11-4.

Table 11-4. *Options When Splitting a Column by Number of Characters*

Option	Description
Number of Characters	Lets you define the number of characters of data before splitting the column.
Once, As Far Left As Possible	Splits the column once only at the given number of characters from the left.
Once, As Far Right As Possible	Splits the column once only at the given number of characters from the right.
Repeatedly	Splits the column as many times as necessary to cut it into segments every defined number of characters.
Advanced Options ➤ Number of Columns to Split Into	Allows you to set a maximum number of columns into which the data is split in chunks of the given number of characters. Any extra columns are placed in the rightmost column.
Split into Columns	This leaves the number of rows as it is in the dataset and creates new columns for each new element resulting from the split operation.
Split into Rows	Creates a new row for each new element resulting from the split operation and duplicates the existing record as many times as there are split elements.

There are a couple of things to note when splitting columns:

- When splitting by a delimiter, Power BI Desktop makes a good attempt at guessing the maximum number of columns into which the source column must be split. If it gets this wrong (and you can see what its guesstimate is if you expand the Advanced options box), you can override the number here.

- If you select a Custom Delimiter, Power BI Desktop displays a new box in the dialog where you can enter a specific delimiter.

- Not every record has to have the same number of delimiters. Power BI Desktop simply leaves the rightmost column(s) blank if there are fewer split elements for a row.

- Conversely, if you specify fewer columns than actually corresponds to the number of available delimiters, then Power BI removes the data that would have gone in the final columns.

Note   You can only split columns if they are text data. The Split Column button remains grayed out if your intention is to try to split a date or numeric column.

Splitting Columns by Character Switch

The final variant on this theme is when the text inside a column switches case–or when text characters switch to numbers (and vice versa).

Options for this type of operation are given in Table 11-5.

Table 11-5.   *Options When Splitting a Column by Number of Characters*

Option	Description
By Lowercase to Uppercase	Any uppercase character immediately following a lowercase character starts a new column.
By Uppercase to Lowercase	Any lowercase character immediately following an uppercase character starts a new column.
By Digit to Non-Digit	Any numeric character immediately following an alphabetical character starts a new column.
By Non-Digit to Digit	Any alphabetical character immediately following a numeric character starts a new column.

Merging Columns

You may be feeling a certain sense of déjà vu when you read the title of this section. After all, we saw how to merge columns (i.e., how to fuse the data from several columns into a single, wider column) in Chapter 8, did we not?

Yes, we did indeed. However, this is not the only time in this chapter that you will see something that you have tried previously. This is because Power Query repeats several of the options that are in the Transform ribbon in the Add Column ribbon. While these functions all work in much the same way, there is one essential difference. If you select an option from the Transform ribbon, then the column(s) that you selected is *modified*. If you select a similar option from the Add Column ribbon, then the original column(s) will not be altered, but a *new column* is added containing the results of the data transformation.

Merging columns is a case in point. Now, as I went into detail as to how to execute this kind of data transformation in the previous chapter, I will not describe it all over again here. Suffice it to say, if you Ctrl-click the headings of two or more columns and then click Merge Columns in the Add Column ribbon, you will still see the data from the selected columns concatenated into a single column. However, this time the original columns *remain* in the dataset. The new column is named Merged, exactly as was the case for the first of the columns that you selected when merging columns using the Transform ribbon.

The following are other functions that can either overwrite the data in existing columns *or* display the result as a new column:

- *Format*: Trims or changes the capitalization of text

- *Extract*: Takes part of a column and creates another column from this data

- *Parse*: Adds a column containing the source column data as JSON or XML strings

- *Statistics*: Creates a new column of aggregated numeric values

- *Standard*: Creates a new column of calculated numeric values

- *Scientific*: Creates a new column by applying certain kinds of math operations to the values in a column

- *Trigonometry*: Creates a new column by applying certain kinds of trigonometric operations to the values in a column

- *Rounding*: Creates a new column by rounding the values in a column

- *Information*: Creates a new column indicating arithmetical information about the values in a column

- *Date*: Creates a new column by extracting date elements from the values in a date column

- *Time*: Creates a new column by extracting time elements from the values in a time or date/time column

- *Duration*: Creates a new column by calculating the duration between two dates or date/times

When transforming data, the art is to decide whether you want or need to keep the original column before applying one of these functions. Yet, once again, it is not really fundamental if you later decide that you made an incorrect decision, as you can always backtrack. Alternatively, you can always decide to insert new columns as a matter of principle and delete any columns that you really do not need at a later stage in the data transformation process.

Creating Columns from Examples

Creating your own columns can be a little scary if you have not had much experience with Excel or Power Pivot formulas, so the Power BI Desktop development team has tried to make your life easier by adding another way to create custom columns. Instead of referring to columns by the column name (and having to handle square brackets and other peculiar characters), you can build a new column by using the actual data in a row. This way, Power Query takes the data as the model–or template–for the transformation to be applied to each row.

The following steps show an example of how to do this:

1. Load the `C:\PowerBiDesktopSamples\Example1.pbix` sample file.

2. Click Transform Data.

3. In the Add Column ribbon, click Column From Examples. A new kind of formula bar will appear above the data. It will look like Figure 11-10. At the same time, a new, empty column will be created at the right of the existing data.

Figure 11-10. *Creating a column from examples*

4. Double-click inside the new column on the right. A list of data from each field will be displayed, as shown in Figure 11-11.

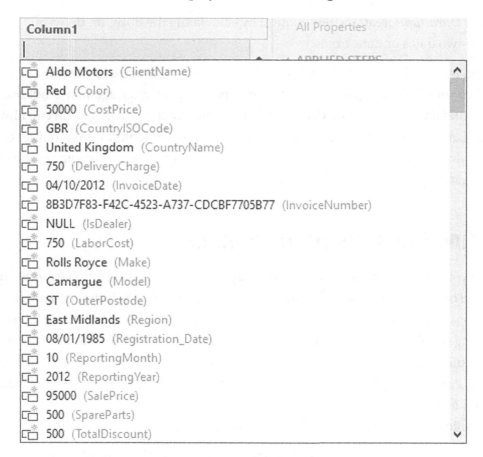

Figure 11-11. *Displaying the data from a row when creating a column from examples*

5. Double-click Red to select the data from the Color column.

6. Enter a space, a hyphen, and a space, and then type **Camargue** (this is the name of the model for this row).

7. Click OK in the formula bar at the top.

Power BI Desktop will add a new column containing the contents of the color column, a separator, and the contents of the model column. Inserted Merged Column will be added as a new step in the Applied Steps list.

Note   In the pop-up menu for the Column From Examples button, you can choose to take all existing columns as the basis for the example or only any columns that you have previously selected.

As you can see from this short example, creating columns by example lets you use the data from a column rather than the column name to transform data.

Tip   If you select Column From Examples ➤ From Selection, then you will only see data from the selected columns when you double-click inside the new column to see samples of data as you did in step 4 of this example.

Adding Conditional Columns

Not all additional columns are a simple extraction or concatenation of existing data. There will be times when you will want to apply some simple conditions that define the contents of a new column. This is where Power BI Desktop's Conditional Column function comes into its own.

Conditional Columns are probably best understood with the aid of a practical example. So let's suppose that you want to add a column that contains a comment on the type of buyer for Brilliant British Cars' products. Here is how you can do this:

1. Load the C:\PowerBiDesktopSamples\Example1.pbix sample file.

2. Click Transform Data.

3. In the Add Column ribbon, click Conditional Column. The Add Conditional Column dialog will appear.

4. Enter **BuyerType** in the "New column name" field.

5. Select Make as the column name.

6. Leave Equals as the operator.

7. Enter **Rolls Royce** as the value.

8. Enter **Posh** as the output.

9.   Click Add Clause.

10.   Select Make as the column name, leave Equals as the operator, enter **Bentley** as the value, and add **Classy** as the output.

11.   Enter **Bling** in the Else field. The dialog will look like Figure 11-12.

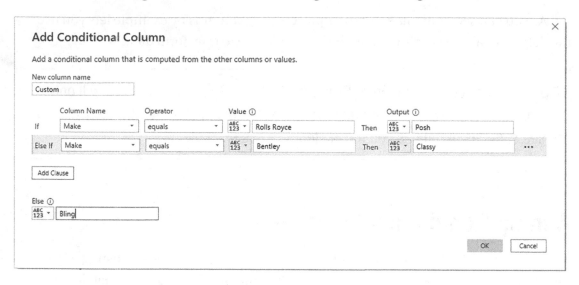

Figure 11-12. *The Add Conditional Column dialog*

12.   Click OK. The rule is applied to every row in the source data, and the new column will be added containing either Posh, Classy, or Bling, depending on the make for each record. Added Conditional Columns will appear as the new step in the Applied Steps list.

As you can see from the Add Conditional Column dialog, it has a range of options that you can tweak when defining the logic for the data matching. These options are outlined in Table 11-6.

Table 11-6. *Custom Column Operators*

Operator	Description
Equals	Sets the text that must match the contents of the selected field for the output to be applied
Does Not Equal	Sets the text that must not match the contents of the selected field for the output to be applied
Begins With	Sets the text at the left of the selected field for the output to be applied
Does Not Begin With	Sets the text that must not appear at the left of the selected field for the output to be applied
Ends With	Sets the text at the right of the selected field for the output to be applied
Does Not End With	Sets the text that must not appear at the right of the selected field for the output to be applied
Contains	Sets the text that can appear anywhere in the selected field for the output to be applied
Does Not Contain	Sets the text that cannot appear anywhere in the selected field for the output to be applied

It is also worth noting that the comparison value, the output, and the alternative output can be values (as was the case in this example), columns, or parameters (which you will learn about in Chapter 13). If you want to remove a rule, simply click the ellipses at the right of the required rule and select Delete.

Tip   Should you wish to alter the order of the rules in the Add Conditional Column dialog, all you have to do is click the ellipses at the right of the selected rule and select Move Up or Move Down from the pop-up menu.

Index Columns

An index column is a new column that numbers every record in the table sequentially. This numbering scheme applies to the table, because it is currently sorted and begins at zero. There are many situations where an index column can be useful. The following are some examples:

- Reapply a previous sort order.

- Create a unique reference for every record.

- Prepare a recordset for use as a dimension table in a Power BI Desktop data model. In cases like this, the index column becomes what dimensional modelers call a *surrogate key*.

This list is not intended to be exhaustive in any way; you will almost certainly find other uses as you work with Power BI Desktop. Whatever the need, here is how to add an index column:

1. In the Add Column ribbon, click Index Column. The new, sequentially numbered column is added at the right of the table, and Added Index is added to the Applied Steps list.

2. Scroll to the right of the table and rename the index column; it is currently named Index.

You have a fairly free hand when it comes to deciding how to begin numbering an index column. The choices are as follows:

- Start at 0 and increment by a value of 1 for each row.

- Start at 1 and increment by a value of 1 for each row.

- Start at any number and increase by any number.

As you saw in step 1, the default is for Power BI Desktop Query to begin numbering rows at 0. However, you can choose another option by clicking the small triangle to the right of the Add Index Column button. This displays a menu with the three options outlined. You can see this menu in Figure 11-13.

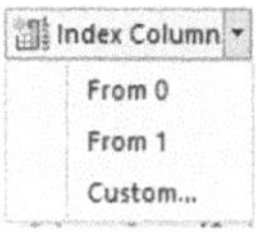

Figure 11-13. *The Add Index Column menu*

Selecting the third option, Custom, displays the dialog that you see in Figure 11-14.

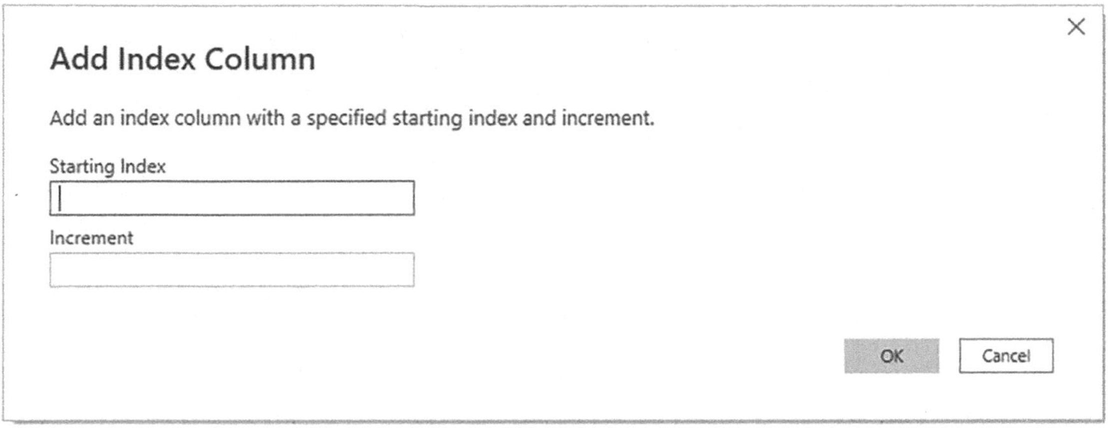

×

Add Index Column

Add an index column with a specified starting index and increment.

Starting Index

Increment

OK   Cancel

Figure 11-14. *The Add Index Column dialog*

This dialog lets you specify the start number for the first row in the dataset as well as the increment that is added for each record.

Conclusion

In this chapter, you learned some essential techniques that you can use to transform datasets. This began with filling down missing values and adding text to existing data. You also saw how to merge the contents of columns.

Finally, you saw a series of techniques that help you to add new columns based on the data in existing columns. These range from extracting parts of a column's data or even deducing different data that is added to a new column using simple logic.

It is now time to see how you can parse complex data types to add them to a dataset. You will even learn how to load multiple identically structured files in a single query and extend the core capabilities of Power Query by adding R and Python scripts. All of this will be the subject of the next chapter.

CHAPTER 12

Complex Data Structures

Not all data loads are a matter of simply establishing a connection to the source and applying transformations to the source data that is, fortunately, already laid out in neatly structured tables. Sometimes, you may want to "push the envelope" when loading data and prepare more complex source data structures for use in your Power BI dashboards. By this, I mean that the source data is not initially in a ready-to-use tabular format and that some restructuring of the data is required to prepare a table for use.

To solve these kinds of challenges, this chapter will explain to you how to

- Add multiple identical files from a source folder

- Select the identical source files to load from a source folder

- Load simple JSON structures from a source file containing JSON data

- Parse a column containing JSON data in a source file

- Parse a column containing XML data in a source file

- Load complex XML files—and select the elements to use

- Convert columns to lists for use in complex load routines

- Apply Python scripts to modify the data

- Apply R scripts to modify the data

Finally—and purely to complete the overall overview of Power Query and its capabilities—I will mention how to

- Reuse recently used queries

- Modify the list of recently used queries

- Export data from Power Query

As ever, all of these transformations will be carried out using Power Query accessed from Power BI Desktop.

347

© Adam Aspin 2022
A. Aspin, *Pro Data Mashup for Power BI*, https://doi.org/10.1007/978-1-4842-8578-7_12

Adding Multiple Files from a Source Folder

Now let's consider an interesting data ingestion challenge. You have been sent a collection of files, possibly downloaded from an FTP site or received by email, and you have placed them all into a specific directory. However, you do not want to have to carry out the process that you saw in Chapter 2 and load files one by one if there are several hundred files—and then append all these files individually to create a final composite table of data (as you saw in Chapter 9).

Note   Power Query can only load multiple files if all the files are rigorously identical. This means ensuring that all the columns are in the same order in each file and have the same names.

Here is a much more efficient method to achieve this objective:

1. Create a new Power BI Desktop file.

2. In the Power BI Desktop Home ribbon, click Transform data to open Power Query (unless you are running Power Query already).

3. Click New Source More ➤ File. Then select Folder to the right of the dialog.

4. Click Connect. The Folder dialog is displayed.

5. Click the Browse button and navigate to the folder that contains the files to load. In this example, it is C:\PowerBiDesktopSamples\ MultipleIdenticalFiles. You can also paste in, or enter, the folder path if you prefer. The Folder dialog will look like Figure 12-1.

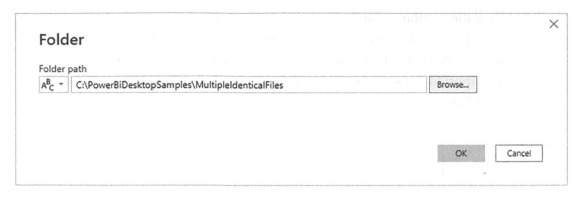

Figure 12-1. *The Folder dialog*

6. Click OK. The file list window opens. The contents of the folder
   and all subfolders are listed in a tabular format, as shown in
   Figure 12-2.

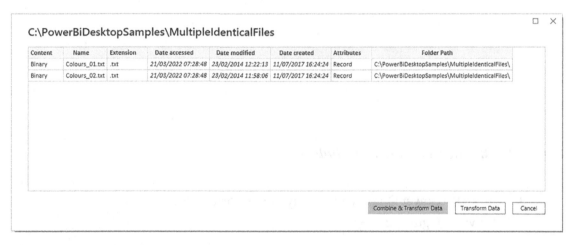

Figure 12-2. *The folder contents in Power BI Desktop*

7. Click the Combine and Transform data button. The Combine
    Files dialog will appear, as shown in Figure 12-3. Here, you can
    select which of the files in the folder is the *model* for the files to
    be imported. This can be a specific file or simply the first file in
    the folder.

Figure 12-3. *The Combine Files dialog*

8. Click OK. Power Query will display the imported data. This is
    shown in Figure 12-4.

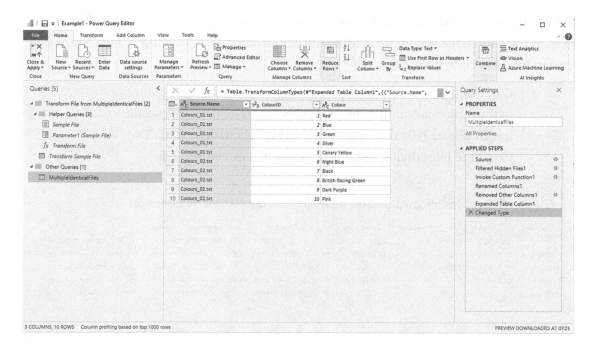

Figure 12-4. *Data loaded from a folder*

9. Remove the automatically generated Source Name column that indicates the source for each record (unless you need it for auditing the data).

10. Click Close & Apply. The data from all the source files will be loaded into the Power BI Desktop data model.

Note   The other options in the Combine Files dialog are explained in Chapter 2.

Filtering Source Files in a Folder

There will be times when you want to import only a *subset* of the files from a folder. Perhaps the files are not identical or maybe you simply do not need some of the available files in the source directory. Whatever the reason, here is a way to get Power BI Desktop to do the work of trawling through the directory and *only* loading files that correspond

to a file name specification you have indicated. In other words, Power Query allows you
to filter the source file set before loading the actual data. In this example, I will show you
how to load multiple Excel files:

1. Carry out the preceding steps 1 through 6 to display the contents
    of the folder containing the files you wish to load. In this scenario,
    it is `C:\PowerBiDesktopSamples\MultipleNonIdentical`. The
    Combine Files dialog will appear as shown in Figure 12-5.

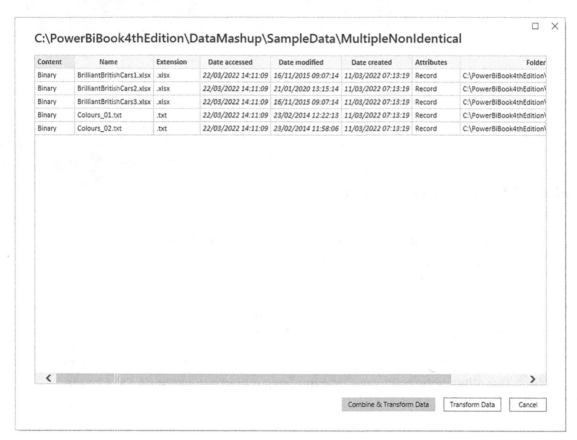

C:\PowerBiBook4thEdition\DataMashup\SampleData\MultipleNonIdentical

Content	Name	Extension	Date accessed	Date modified	Date created	Attributes	Folder
Binary	BrilliantBritishCars1.xlsx	.xlsx	22/03/2022 14:11:09	16/11/2015 09:07:14	11/03/2022 07:13:19	Record	C:\PowerBiBook4thEdition\
Binary	BrilliantBritishCars2.xlsx	.xlsx	22/03/2022 14:11:09	21/01/2020 13:15:14	11/03/2022 07:13:19	Record	C:\PowerBiBook4thEdition\
Binary	BrilliantBritishCars3.xlsx	.xlsx	22/03/2022 14:11:09	16/11/2015 09:07:14	11/03/2022 07:13:19	Record	C:\PowerBiBook4thEdition\
Binary	Colours_01.txt	.txt	22/03/2022 14:11:09	23/02/2014 12:22:13	11/03/2022 07:13:19	Record	C:\PowerBiBook4thEdition\
Binary	Colours_02.txt	.txt	22/03/2022 14:11:09	23/02/2014 11:58:06	11/03/2022 07:13:19	Record	C:\PowerBiBook4thEdition\

Combine & Transform Data   Transform Data   Cancel

Figure 12-5. *The Combine Files dialog*

2. Click Transform Data. Power Query will open and display the list
    of files in the directory and many of their attributes. You can see
    an example of this in Figure 12-6.

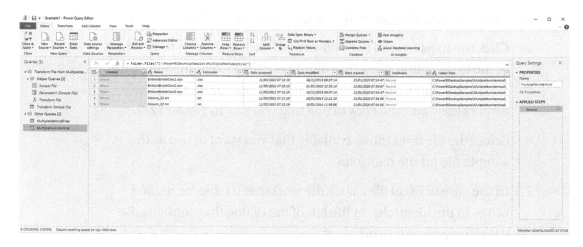

Figure 12-6. *Displaying file information when loading multiple files*

3. As you want to load only Excel files, and avoid files of any other type, click the filter pop-up menu for the column title Extension and uncheck all elements *except* .xlsx. This is shown in Figure 12-7.

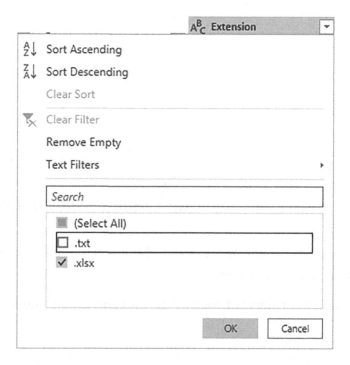

Figure 12-7. *Filtering file types when loading multiple identical files*

4. Click OK. You will now only see the Excel files in Power Query.

5. Click the Expand icon (two downward-facing arrows) to the right of the first column title; this column is called Content, and every row in the column contains the word *Binary*. Power Query will display the Combine Files dialog that you saw in Figure 12-3.

6. Select the file from those available that you want to use as the sample file for the data load.

7. In the case of Excel files, click the worksheet, table, or named range in the hierarchy on the left of the dialog that contains the data that you wish to load.

8. Click Skip files with errors. This time, the dialog will look like Figure 12-8 (for Excel files).

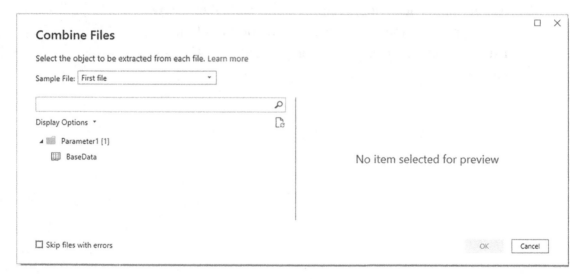

Figure 12-8. *Selecting the source data when loading multiple Excel files*

9. Select the worksheet or table to be used from all the similar source files (BaseData in this example).

10. Click OK. Power Query will load all the files and display the result.

The contents of all the source files are now loaded into Power Query and can be transformed and used like any other dataset. This might involve removing superfluous header rows (as described in the next but one section) or removing the column that contains the source file name.

What is more, if ever you add more files to the source directory and then click Refresh in the Home ribbon, *all* the source files that match the filter selection are reloaded, including any new files added to the specified directory that match the filter criteria.

There are a few things to note here:

- If you click Combine & Transform at step 4, Power Query will complete the data load for all files.

- If you check the check box Skip files with errors, only files that map to the initial file model–and that do not cause any issues when loading– will be loaded.

- When loading multiple Excel files, you need to be aware that the data sources (whether they are worksheets, named ranges, or tables) *must* have the same name in all the source files or the data will not be loaded.

- If you are loading *text* files, you will see, in step 8, a Combine Files dialog similar to the one shown in Figure 12-3.

Displaying and Filtering File Attributes

When you display the contents of a folder in Power Query, you see a set of file attributes that you can use to filter data. These cover basic elements such as

- File name

- File extension

- Folder path

- Date created

- Date last accessed

- Date modified

However, there are many more attributes that are available to describe files that you can access simply by displaying them in Power Query. Here is how you can do this:

1. Carry out steps 1 and 2 from the previous section.

2. Display the available attributes by clicking the Expand icon (the double-headed arrow) at the right of the Attributes column. The list of available attributes will be displayed, as shown in Figure 12-9.

Figure 12-9. Selecting the source data when loading multiple Excel files

3. Select the attributes that you want to display from the list and click OK.

Each attribute will appear as a new column in Power Query. You can now filter on the columns to select files based on the expanded list of attributes.

Note   You can also filter on directories, dates, or any of the file information that is displayed. Simply apply the filtering techniques that you learned in Chapter 8.

The List Tools Transform Ribbon

Power BI Desktop considers some data to be lists, not tables of data. It handles lists slightly differently and displays a specific ribbon to modify list data. The List Tools Transform ribbon is explained in Figure 12-10 and Table 12-1.

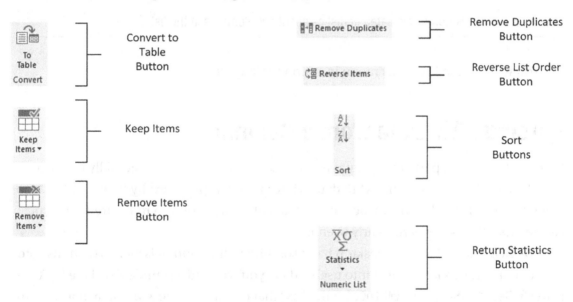

Figure 12-10. *The List Tools Transform ribbon*

Table 12-1. *The List Tools Transform Ribbon Options*

Option	Description
To Table	Converts the list to a table structure
Keep Items	Allows you to keep a number of items from the top or bottom of the list or a range of items from the list
Remove Items	Allows you to remove a number of items from the top or bottom of the list or a range of items from the list
Remove Duplicates	Removes any duplicates from the list
Reverse Items	Reverses the list order
Sort	Sorts the list lowest to highest or highest to lowest
Statistics	Returns calculated statistics about the elements in the list

You will see the List Tools menu in action in Chapter 13.

Parsing XML Data from a Column

Some data sources, particularly database sources, include XML data actually inside a field. The problem here is that XML data is interpreted as plain text by Power BI Desktop when the data is loaded. If you look at the AvailableColors column in Figure 12-11, you can see that this is not particularly useful.

So once again, Power BI Desktop has a solution to this kind of issue. To demonstrate how to convert this kind of text into usable data, you will find a sample Excel file (C:\ PowerBiDesktopSamples\XMLInColumn.xlsx) that contains some XML data in a column. Proceed as follows:

1. In a new, blank Power BI Desktop file, click Edit Queries to switch to Power Query.

2. In the Home ribbon, select New Source ➤ Excel.

3. Select the Excel file C:\PowerBiDesktopSamples\ XMLInColumn.xlsx.

4. Select the only worksheet in this file. It is named Sales.

5. Scroll to the right of the dataset and take a look at the last column: AvailableColors. The Navigator dialog looks like Figure 12-11.

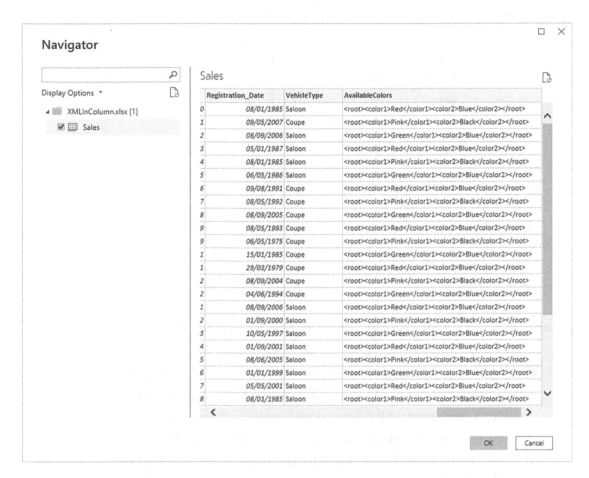

Figure 12-11. *A column containing XML*

6. Click OK to connect to the source file.

7. Select the AvailableColors column at the right of the dataset.

8. In the Add Column ribbon, click Parse ➤ XML. A new column will be added to the right. It will look like Figure 12-12 and will have the title XML.

A^B_C AvailableColors	ABC 123 XML
<root><color1>Red</color1><color2>Blue</color2></root>	Table
<root><color1>Pink</color1><color2>Black</color2></root>	Table
<root><color1>Green</color1><color2>Blue</color2></root>	Table
<root><color1>Red</color1><color2>Blue</color2></root>	Table
<root><color1>Pink</color1><color2>Black</color2></root>	Table
<root><color1>Green</color1><color2>Blue</color2></root>	Table
<root><color1>Red</color1><color2>Blue</color2></root>	Table
<root><color1>Pink</color1><color2>Black</color2></root>	Table
<root><color1>Green</color1><color2>Blue</color2></root>	Table
<root><color1>Red</color1><color2>Blue</color2></root>	Table
<root><color1>Pink</color1><color2>Black</color2></root>	Table
<root><color1>Green</color1><color2>Blue</color2></root>	Table
<root><color1>Red</color1><color2>Blue</color2></root>	Table
<root><color1>Pink</color1><color2>Black</color2></root>	Table
<root><color1>Green</color1><color2>Blue</color2></root>	Table

Figure 12-12.   *An XML column converted to a table column*

9. Click the Expand icon to the right of the XML column title and uncheck "Use original column name as prefix" in the pop-up dialog. Ensure that all the columns are selected and click OK. Two new columns (or, indeed, as many new columns as there are XML data elements) will appear at the right of the dataset. Power Query will look like Figure 12-13.

Figure 12-13. *XML data expanded into new columns*

10. Remove the column containing the initial XML data by selecting the column that contains the original XML and clicking Remove Columns in the context menu.

Using this technique, you can now extract the XML data that is in source datasets and use it to extend the original source data.

Parsing JSON Data from a Column

Sometimes, you may encounter data containing JSON in a field too. The technique to extract this data from the field inside the dataset and convert it to columns is virtually identical to the approach that you saw in the previous section for XML data.

Given that the approach is so similar—and is not far removed from what you saw previously when importing XML files—I will only provide a screenshot for the final result of the process. Here, you will be able to see the source JSON as well as the columns of data that were extracted from the JSON and added to the dataset.

1. In a new, blank Power BI Desktop file, load the data from the Excel file C:\PowerBiDesktopSamples\JSONInColumn.xlsx and switch to Power Query. Select the only worksheet in this file: Sales.

2. Scroll to the right of the dataset and select the last column: AvailableColors.

3. In the Add Column ribbon, click Parse ➤ JSON. A new column will be added to the right and will have the title JSON.

4. Click the Expand icon to the right of the JSON column title and uncheck "Use original column name as prefix" in the pop-up dialog. Ensure that all the columns are selected and click OK. Two new columns (or, indeed, as many new columns as there are JSON data elements) will appear at the right of the dataset. Power Query will look like Figure 12-14.

A^B_C AvailableColors	▼	ABC 123 Color1	▼	ABC 123 Color2	▼
{"Color1":"Red", "Color2":"Blue"}		Red		Blue	
{"Color1":"Red", "Color2":"Blue"}		Red		Blue	
{"Color1":"Red", "Color2":"Blue"}		Red		Blue	
{"Color1":"Red", "Color2":"Blue"}		Red		Blue	
{"Color1":"Red", "Color2":"Blue"}		Red		Blue	
{"Color1":"Red", "Color2":"Blue"}		Red		Blue	
{"Color1":"Red", "Color2":"Blue"}		Red		Blue	
{"Color1":"Red", "Color2":"Blue"}		Red		Blue	
{"Color1":"Red", "Color2":"Blue"}		Red		Blue	
{"Color1":"Red", "Color2":"Blue"}		Red		Blue	
{"Color1":"Red", "Color2":"Blue"}		Red		Blue	
{"Color1":"Red", "Color2":"Blue"}		Red		Blue	
{"Color1":"Red", "Color2":"Blue"}		Red		Blue	
{"Color1":"Red", "Color2":"Blue"}		Red		Blue	
{"Color1":"Red", "Color2":"Blue"}		Red		Blue	
{"Color1":"Red", "Color2":"Blue"}		Red		Blue	
{"Color1":"Red", "Color2":"Blue"}		Red		Blue	

Figure 12-14. *JSON data expanded into new columns*

5.   Delete the column containing the initial JSON data.

Admittedly, the structure of the JSON data in this example is extremely simple. Real-world JSON data could be much more complex. However, you now have a starting point upon which you can build when parsing JSON data that is stored in a column of a dataset.

Complex JSON Files

JSON files are not always structured as simplistically as the colors.json file that you saw in the previous section. Indeed, JSON files can contain many sublevels of data, structured into separated *nodes*. Each node may contain multiple data elements grouped together in a logical way. Often, you will want to select "sublevels" of data from the source file—or perhaps only select some sublevel elements and not others.

This section shows you how the data elements from a complex JSON structure appear after data load. Specifically, the sample source data file contains a "root" level that displays core data such as the invoice number, sale date, and sale price (among other elements) and three "sublevels" that contain information on

- The vehicle

- The finance data

- The customer

Note   If you want to get an idea of what a complex JSON file containing several nested nodes looks like, then simply open the file `C:\PowerBIDesktopSamples\CarSalesJSON_Complex.json` in a text editor.

---

1. In Power Query, click New Source ➤ More ➤ File.

2. Select JSON on the right and click Connect.

3. Choose the file C:\PowerBiDesktopSamples\CarSalesJSON_Complex.Json.

4. Click Open.

You simply need to be aware that Power Query creates column titles from complex JSON files that reflect the hierarchy of data in the source file. So you will see column names like Sales.VehicleModel that indicate that the VehicleModel field was nested inside the Sales node of the source JSON. You can rename—or remove—any columns, of course. You can see the initial load status of a complex JSON file in Figure 12-15.

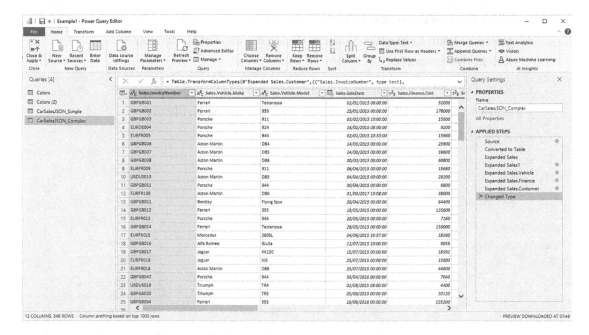

Figure 12-15. *A complex JSON file after initial load*

Tip   It is a good idea to click the Load more link in the Expand pop-up menu when you are identifying the nested data in a JSON node. This will force Power Query to scan a larger number of records and return, potentially, a more complete list of nested fields.

If you want to understand exactly how Power Query transformed the data, you can click through the "Applied Steps" to gain more of an understanding of the steps Power Query has applied when extracting complex JSON.

Complex XML Files

As is the case with JSON files, XML files can comprise complex nested structures of many sublevels of data, grouped into separate nodes. In the case of complex XML files, however, Power Query needs you to apply a few more steps to the import process.

1.   In the Power BI Desktop ribbon, click Get Data ➤ File and select XML in the list of file sources on the right of the Get Data dialog.

2. Click Connect and navigate to the folder containing the XML file that you want to load (`C:\PowerBIDesktopSamples\ComplexXML.Xml` in this example).

3. Select the XML node named Sales. You can see in the Navigator–shown in Figure 12-16–that all subnodes of the XML file are interpreted as tables by Power Query.

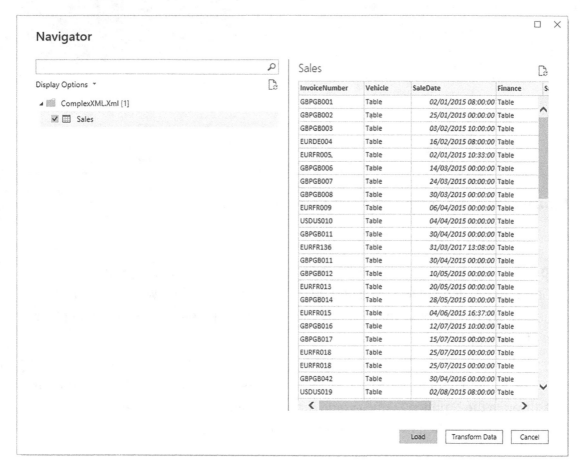

Figure 12-16. *Navigator view of a complex XML file*

4. Click Transform Data. Power Query window will appear.

5. Click Open. The Query Editor window will appear and automatically display the tables that contain sublevels of data. You can see this in Figure 12-17.

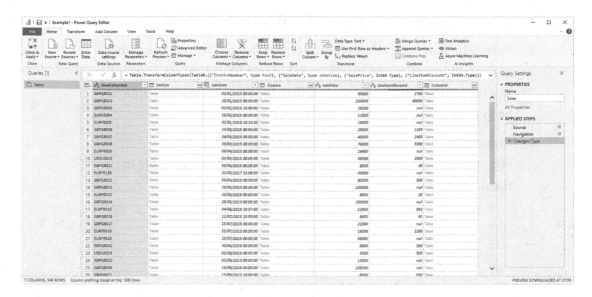

Figure 12-17. *Viewing the structure of an XML file*

6. Select the Vehicle column and click the Expand icon at the right of the column title. The list of available elements that are "nested" at a lower level inside the source XML will appear. You can see this in Figure 12-18.

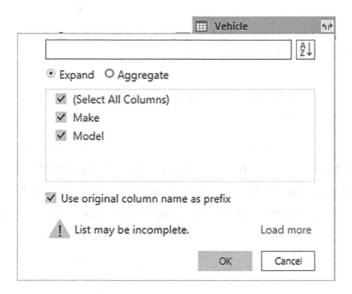

Figure 12-18. *Nested elements in an XML file*

7. Click OK. The new columns will be added to the data table.

8. Select the Finance column and click the Expand icon at the right of the column title. The list of available elements that are "nested" at a lower level inside the source XML for this column will appear. Select only the Cost column and click OK.

9. Remove the Customer column. Power Query will look like Figure 12-19, where all the required columns are now visible in the data table.

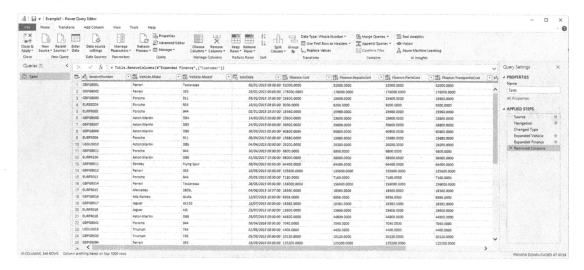

Figure 12-19. *An XML file after parsing*

10. Click the Apply Changes button at the top of the Power BI Desktop window.

Tip   It is a good idea to click the Load more link in the Expand pop-up menu when you are identifying the nested data in an XML node. This will force Power Query to scan a larger number of records and return, potentially, a more complete list of nested fields.

This approach allows you to be extremely selective about the data that you load from an XML file. You can choose to include any column at any level from the source structure. As you saw, you can select—or ignore—entire sublevels of nested data extremely easily.

This section was only a simple introduction to parsing complex XML files. As this particular data structure can contain multiple sublevels of data and can mix data and sublevels in each "node" of the XML file, the source data structure can be extremely complex. Fortunately, the techniques that you just learned can be extended to handle any level of XML complexity and help you tame the most daunting data structures.

Note   It is important to "flatten" the source data so that all the sublevels (or nodes if you prefer) are removed and the data that they contain is displayed as a simple column in the query.

Python and R Scripts

As complete and powerful as Power Query may be, it does not provide solutions to every conceivable data load and modification challenge that you may face. Occasionally, you may find yourself needing to call in the "big guns" of the data transformation world and invoke external languages such as Python and/or R to perform complex functions that Power Query cannot do–or, at least, cannot do easily.

Of course, this implies that you need to know some R or Python when you feel the need to apply them to a data transformation process. Or it may be that you are already proficient in these languages and prefer to use the knowledge that you have already acquired. Alternatively, you may already have (or have been given) R or Python scripts and simply want to use them in a Power Query process.

Whatever the reason, Power Query can be extended with the almost limitless capabilities of R and Python extremely easily. As learning these languages is way beyond the scope of this book, the idea in the following two sections is simply to show you how R and Python can be *integrated* into Power Query. The actual R and Python that you use is up to you. The scripts that you add can be as simple or as complex as you require.

Using Python Scripts to Modify Data

To compensate for the fact that Power Query does not have regular expressions (a kind of turbocharged search and–if required–replace function), it is easy to call on Python to overcome this particular lacuna.

Suppose that you want to apply a regular expression to a Power Query process. Let's imagine that you want to find every make that starts with "Rolls" and replace it with "Roller!".

1. Open the Power BI Desktop file Example1.pbix and open Power Query.

2. In the Transform ribbon, click the Py (Run Python script) button. The Run Python script dialog will be displayed.

3. Enter the following snippet of Python:

```
import pandas as pd
pattern = "^Rolls.*"
dataset['TestOutput'] = dataset["Make"].str.replace(pattern,
"Roller!")
```

The dialog will look like Figure 12-20.

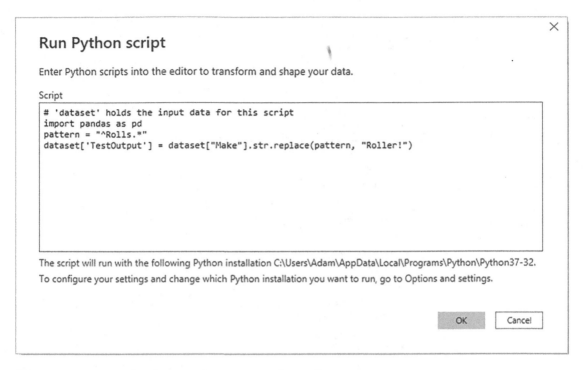

Figure 12-20. *A Python script to transform data*

4. Click OK. You may be asked to ignore privacy checks. The Python script will then run, and the output will look like Figure 12-21.

Figure 12-21. *The result of a Python script*

5. Expand the Value column by clicking the double-headed arrow at the right of the column name and select the columns that you want to use. You can see this in Figure 12-22.

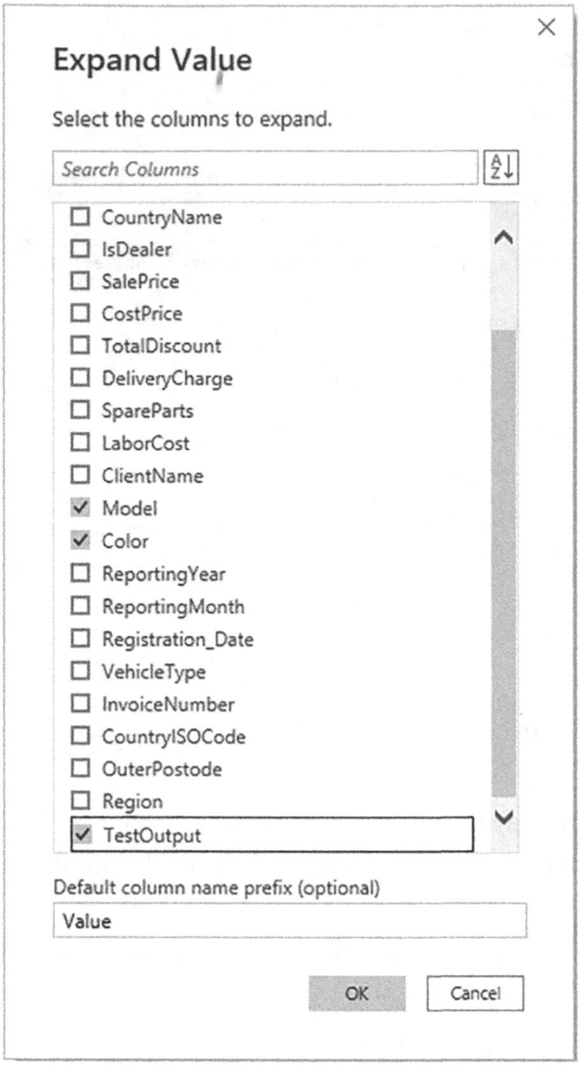

Figure 12-22. *Expanding the table that results from a Python script*

6. Click OK. You will see the result of the Python script as the new
    TestOutput column as shown in Figure 12-23.

	A^BC Name	ABC 123 Value.Make	ABC 123 Value.Model	ABC 123 Value.Color	ABC 123 Value.TestOutput
1	dataset	Rolls Royce	Camargue	Red	Roller!
2	dataset	Aston Martin	DBS	Blue	Aston Martin
3	dataset	Rolls Royce	Silver Ghost	Green	Roller!
4	dataset	Rolls Royce	Silver Ghost	Blue	Roller!
5	dataset	Rolls Royce	Camargue	Canary Yellow	Roller!
6	dataset	Rolls Royce	Camargue	British Racing Green	Roller!
7	dataset	Aston Martin	DBS	Dark Purple	Aston Martin
8	dataset	Aston Martin	DB7	Red	Aston Martin
9	dataset	Aston Martin	DB9	Blue	Aston Martin
10	dataset	Aston Martin	DB9	Silver	Aston Martin
11	dataset	Aston Martin	DB4	Night Blue	Aston Martin
12	dataset	Aston Martin	Vantage	Canary Yellow	Aston Martin
13	dataset	Aston Martin	Vanquish	Night Blue	Aston Martin
14	dataset	Aston Martin	Rapide	Black	Aston Martin
15	dataset	Aston Martin	Zagato	British Racing Green	Aston Martin
16	dataset	Rolls Royce	Silver Ghost	Canary Yellow	Roller!
17	dataset	Rolls Royce	Wraith	Silver	Roller!
18	dataset	Rolls Royce	Silver Ghost	Green	Roller!
19	dataset	Rolls Royce	Camargue	Blue	Roller!
20	dataset	Rolls Royce	Silver Shadow	Red	Roller!
21	dataset	Rolls Royce	Silver Seraph	Red	Roller!
22	dataset	Rolls Royce	Silver Ghost	Black	Roller!
23	dataset	Rolls Royce	Silver Shadow	British Racing Green	Roller!

Figure 12-23. *The result of applying a Python script*

You can, of course, remove the Name column that only contains the dataset name to finalize this output. You may also wish to rename the columns.

Note   Regular Expressions are a huge subject, and explaining them is beyond the scope of this book.

Using R Scripts to Modify Data

As an example of an R script, I want to show something that is less a transformation and more a complete import function. This is to give you an idea of how deeply integrated R (and Python) is into Power Query.

One missing built-in function in Power Query is the ability to unzip compressed source files. However, as R can unzip source files, it only takes one line of R script to unzip a CSV file and have it ready for further data transformation. You can see this in action in the following example:

1. Open the Power BI Desktop file Example1.pbix and open Power Query.

2. Click New source ➤ Blank Query.

3. In the View ribbon, click Advanced editor.

4. Enter the following R code as shown in Figure 12-24 (the code snippet is available in the file RCode.txt in the sample files):

```
R.Execute("countrydata<-read.csv(unz(""c:\\PowerBiDesktopSamples\\
Countries.zip"",""Countries.csv""))")
```

The dialog will look like Figure 12-24.

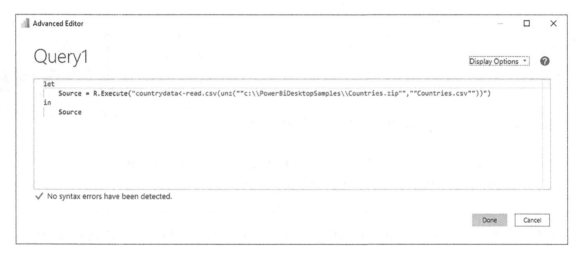

Figure 12-24. *A single line R script*

5. Click Done. You may be asked to confirm permissions to run this kind of code. The output will look like Figure 12-25.

Figure 12-25. *Compressed source data unzipped using R*

6. Expand the Value column by clicking the double-headed arrow at the right of the column.

7. Click OK.

8. Remove the column called Name. You can see the output in Figure 12-26.

⊞▾	ABC 123 CountryID	▾	ABC 123 CountryName	▾	ABC 123 CountryName_Local	▾
1	1		United Kingdom		England	
2	2		France		France	
3	3		USA		United States	
4	4		Germany		Deutschland	
5	5		Spain		Espana	
6	6		Switzerland		Suisse	

Figure 12-26. *An uncompressed CSV file produced by an R script*

I will not explain the R script here. However, it is fairly comprehensible and shows how R and Power Query can be used together to deliver some easy and wide-ranging data load and transformation effects.

Note   You can add R scripts that transform data that is already loaded simply by clicking the Run R script button in the Transform menu (as you did for the Python script in the previous section). This will depend on the kind of transformation that you want to carry out.

You will learn a lot more about the Advanced editor in Chapter 14.

Convert a Column to a List

Sometimes, you will need to use data in a list format. You will see a practical example of this in Chapter 13 when you learn how to parameterize queries. Fortunately, Power Query lets you convert a column to a list really easily:

1. Click Get Data ➤ Excel and connect to the Excel file C:\PowerBiDesktopSamples\BrilliantBritishCars.xlsx.

2. Select the worksheet BaseData and click Transform Data to open Power Query.

3. Select a column to convert to a list by clicking the column header. I will use the column Make in this example.

4. In the Transform ribbon, click Convert to List. Power Query will show the resulting list, as you can see in Figure 12-27.

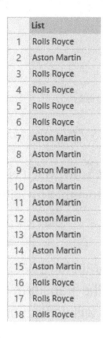

Figure 12-27. *The list resulting from a conversion-to-list operation*

This list can now be used in certain circumstances when carrying out more advanced data transformation processes.

Query Folding

To conclude this chapter, I want to introduce you to a concept that may seem a little abstruse at first sight, but that can prove invaluable in practice. This is *query folding*.

Query folding is an essential—but hidden—aspect of database connections. What it means is that with a little care and attention, you can ensure that all the hard work that underlies any data transformations that you make is carried out by the server that you are connecting to. It is a good idea to understand query folding earlier rather than later–so that you can bear this in mind when writing complex data transformation steps later on.

So what is query folding?

Query folding is the way that Power Query converts (or tries to convert) all your data mashup steps–from selecting tables and columns to many other data mashup approaches–into the SQL that is used by the database that you are connected to. SQL, as you probably know, is the (mostly) generic language used to query databases. So under the covers, Power Query converts anything you define as part of a database data source to SQL. This brings several advantages to the process:

- The process of loading data can use the power of the database (and most of the databases that you can connect to have been around for years and are highly optimized) to work on the source data.

- Data is selected and filtered at source, and only the required data is returned to Power BI Desktop. This avoids bringing back data that is then simply excluded once in Power BI.

- Query folding applies both to imported datasets and data returned using DirectQuery.

The essential data functions that query folding can apply are

- Selecting data

- Filtering data

- Grouping data

- Renaming columns

- Joining source tables in the database

- Applying simple calculations

- Applying simple logic

- Applying basic text manipulation functions

Essentially, you need to know if a query can be folded.

1. Click Get Data ➤ SQL Server database (or any other database that accepts query folding).

2.   Connect to the source database.

3.   Click Transform to open Power Query.

4.   Right–click on a step in the Applied Steps list. The context menu
     will appear as shown in Figure 12-28.

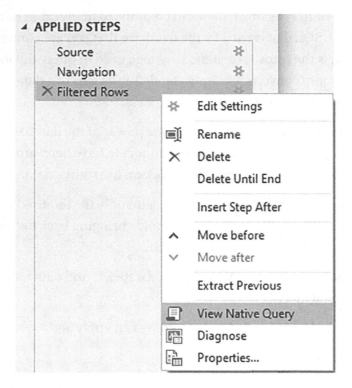

Figure 12-28.   *The Applied Steps pop-up menu*

5.   Click View Native Query. The Native Query dialog will appear as
     shown in Figure 12-29.

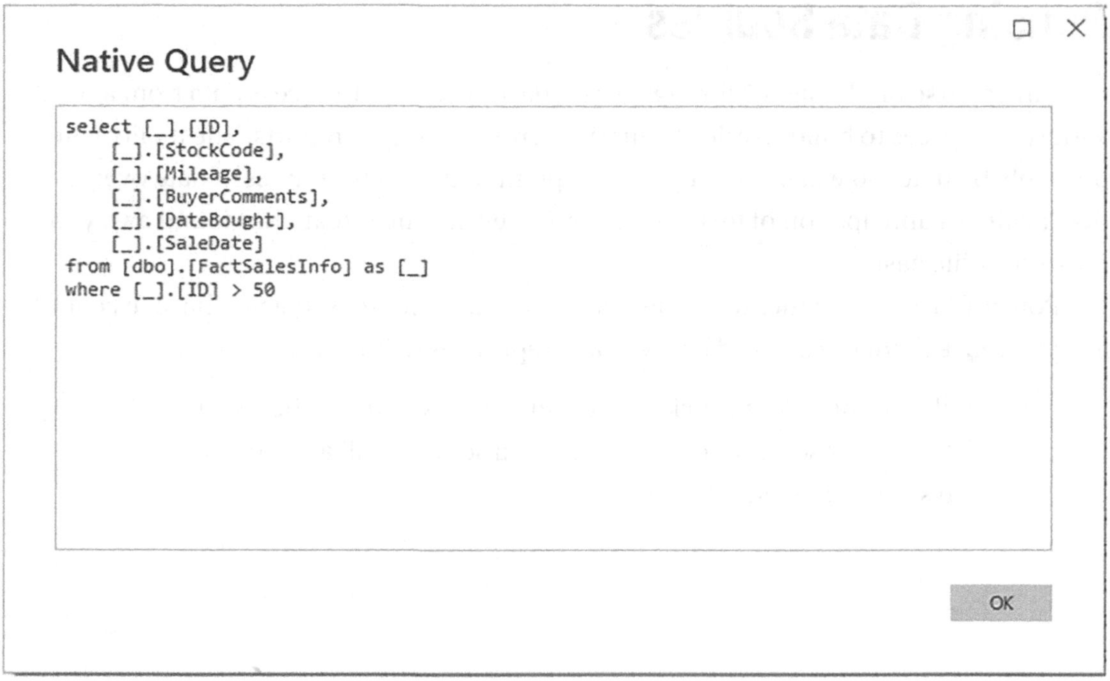

```
select [_].[ID],
    [_].[StockCode],
    [_].[Mileage],
    [_].[BuyerComments],
    [_].[DateBought],
    [_].[SaleDate]
from [dbo].[FactSalesInfo] as [_]
where [_].[ID] > 50
```

OK

Figure 12-29. *The Native Query dialog*

6. Review the SQL and click OK.

Note   If the option View Native Query is grayed out, this means that not all query steps can be folded.

The fact that you can see valid SQL tells you that the query *can* be folded. This means that the SQL will be applied at the source database level. So transformations up to this point will be carried out by the database engine–*not* Power Query.

If a query cannot be folded, you are forcing Power Query to load all the data and then process it in Power Query. This can be–sometimes considerably–slower. In cases like these, you could consider

- Reorganizing queries to add "nonfolding" steps at the end of the query

- Splitting queries to separate out folded steps and then apply "nonfolding" steps

Reusing Data Sources

Over the course of Chapters 2 through 6, you have seen how to access data from a wide variety of sources to build a series of queries across a range of reports. The reality will probably be that you will frequently want to point to the same sources of data over and over again. In anticipation of this, the Power BI development team has found a way to make your life easier.

Power BI Desktop remembers the most recent data sources that you have used and lets you reuse them quickly and easily in any report. Here is how:

1. In the Home ribbon, click the Recent Sources button. The list of the dozen or so most recently used data sources will appear. You can see this in Figure 12-30.

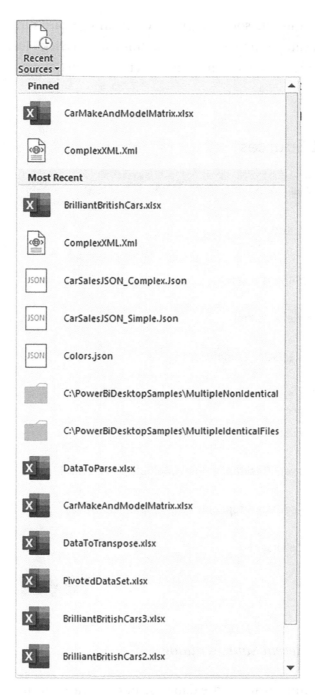

Figure 12-30. *Recently used sources*

2. Click the source that you want to reconnect to, and continue with the data load or connection.

If you cannot see the data source that you want, and you are sure that you have used it recently, then you can scroll to the bottom of this list and click More. Power BI Desktop will display the complete list of recent sources in the Recent Sources dialog that you can see in Figure 12-31.

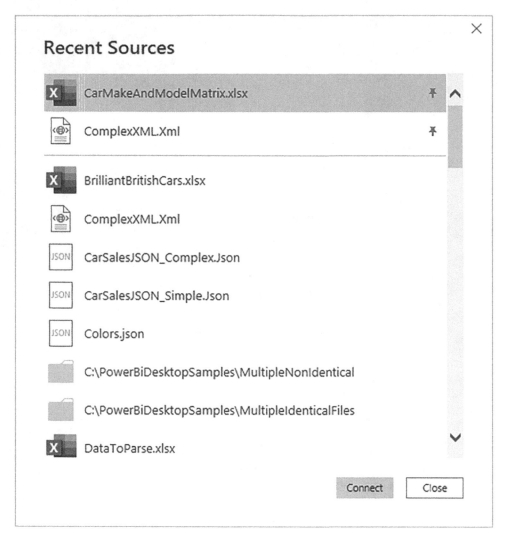

Figure 12-31. *The Recent Sources dialog*

If you are connecting to any of the database or data warehouse sources that allow DirectQuery or Live Connection, you may see a dialog like the one in Figure 12-32.

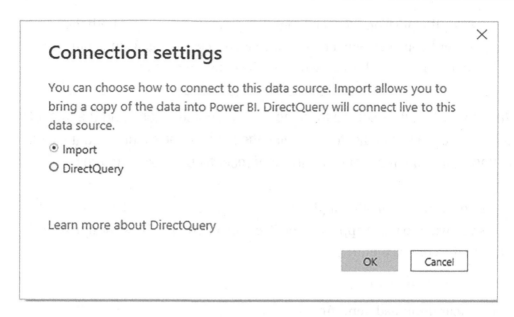

Figure 12-32. Defining connection settings when reusing an existing connection

Here, you can decide whether to load data or use a direct connection, this time with the chosen data source.

It is also worth noting that for database and data warehouse sources, Power Query remembers the server and possibly the database—but not the actual tables used.

Pinning a Data Source

If you look closely at Figure 12-31, you see that the Excel file CarMakeAndModelMatrix. xlsx (among others) is pinned to the top of the Recent Sources dialog. This allows you to make sure that certain data sources are always kept on hand and ready to reuse.

Do the following to pin a data source that you have recently used to the menu and dialog of recent sources:

1. Click the Recent Sources button in the Power BI Desktop
   Home ribbon.

2. Scroll down to the bottom of the menu and click More. The Recent
   Sources dialog will appear.

3. Hover the mouse over a recently used data source. A pin icon will
   appear at the right of the data source name.

4. Click the pin icon. The data source is pinned to the top of both the Recent Sources menu and the Recent Sources dialog. A small pin icon remains visible at the right of the data source name.

Note   To unpin a data source from the Recent Sources menu and the Recent Sources dialog, all you have to do is click the pin icon for a pinned data source. This unpins it and it reappears in the list of recently used data sources.

If you so wish, you can also apply the following options when deciding which elements you want to make appear in the Recent Sources list. These options are also available in the context menu.

- Remove from list

- Clear unpinned items from list

Copying Data from Power Query

Power BI Desktop is designed as a data destination. It does not have any data export functionality as such. You can manually copy data from the Power Query, however. More precisely, you can copy any of the following:

- The data in the query

- A column of data

- A single cell

In all cases, the process is the same:

1. Click the element to copy. This can be

    a. The top-left square of the data grid

    b. A column title

    c. A single cell

2. Right-click and select Copy from the context menu.

You can then paste the data from the clipboard into the destination application.

Note   This process is somewhat limited because you cannot select a range of cells. And you must remember that you are only looking at sample data in Power Query. You can, however, click Close & Apply and then switch to Table view from where you can copy an entire table. This process is described in the companion volume *Pro Power BI Dashboard Creation*.

Conclusion

This chapter pushed your data transformation knowledge with Power Query to a new level by explaining how to deal with multiple file loads of Excel- and text-based data. You then learned ways of handling data from source files that contain complex, nested source structures—specifically JSON and XML files. You also saw how to parse JSON and XML elements from columns contained in other data sources.

Then, you learned how to reuse data sources and manage frequently used data sources to save time. You also learned about extending Power Query by including R and Python scripts. Then you learned the essentials concerning query folding. Finally, you learned how to copy sample data resulting from a data transformation process into other applications.

So now the basic tour of data load and transformation with Power Query is over. It is time to move on to more advanced techniques that you can apply to accelerate and enhance, manage, and structure your data transformation processes. These approaches are the subject of the following two chapters.

CHAPTER 13

Organizing, Managing, and Parameterizing Queries

Producing a robust and efficient data query is not just about finding the appropriate load and transform functions and placing them in the correct sequence. It is also about extending, adjusting, and maintaining the process. This can be either to correct an error once the query is being tested or to adapt a query to new requirements. This chapter will introduce you to some of the techniques that you can apply to handle the various stages of the query life cycle.

Creating an eye-catching dashboard can mean sourcing data from a large range and variety of queries. It may also imply that these queries have to be linked together to create a cascade of data transformations that prepares the core elements of a practical and usable data model collated from multiple sources. It follows that you will therefore need to know how to *manage* the queries that you create to use them efficiently and to keep your queries under control in real-world situations.

However, not all import processes (which are also referred to as dataflows) are rigid and predictable. There will, inevitably, be cases where you also want to shape the data ingestion process depending on aspects of the source data. This can mean parameterizing your queries to allow user interaction or adjusting the dataflow dynamically–which will also be explained in this chapter.

© Adam Aspin 2022
A. Aspin, *Pro Data Mashup for Power BI*, https://doi.org/10.1007/978-1-4842-8578-7_13

Managing the Transformation Process

Pretty nearly all the transformation steps that we have applied so far have been individual elements that can be applied to just about any data table. However, when you are carrying out even a simple data load and transform process, you are likely to want to step through several transformations in order to shape, cleanse, and filter the data to get the result you want. This is where the Power BI Desktop approach is so malleable, because you can apply most data transformation steps to just about any data table. The art consists of placing them in a sequence that can then be reused any time that the data changes to reprocess the new source data and deliver an up-to-date output.

The key to appreciating and managing this process is to get well acquainted with the Applied Steps list in the Query Settings pane. This list contains the details of every step that you applied, in the order in which you applied it. Each step retains the name that Power BI Desktop gave it when it was created, and each can be altered in the following ways:

- Renamed

- Deleted

- Moved (in certain cases)

The even better news is that in many cases, steps can be modified. This way, you are not stuck with the choices that you made initially but have the opportunity of tweaking and improving individual steps in a process. This can avoid your having to rebuild an entire sequence of steps in an ETL routine simply by replacing one element in the ETL process.

In order to experiment with the various ways that you can modify queries, you are going to need some initial data. So to start with, I suggest that you create a query that loads data from the following Excel source file: `C:\PowerBiDesktopSamples\CarSalesDataForQueries.xlsx`. From this source file, select the following tables:

- Clients

- Colors

- Countries

- Invoices

- InvoiceLines

- Stock

Once you have loaded the data, switch to Power Query.

Modifying a Step

How you alter a step will depend on how the original transformation was applied. This becomes second nature after a little practice and will always involve first clicking the step that you wish to modify and then applying a different modification. If you invoke a ribbon option, such as altering the data type, then you change the data type by simply applying another data type directly from the ribbon. If you used an option that displayed a dialog (such as splitting a column, among others), then you can right-click the step in the Applied Steps list and select Edit Settings from the context menu. Alternatively, and if you prefer, you can click the "gear" icon that is displayed to the right of most (but not all) steps to display a dialog where you can adjust the step settings. This dialog will show all the options and settings that you applied initially; in it, you can make any modifications that you consider necessary.

A final possibility that makes it easy to alter the settings for a processing step is to edit the formula that appears in the formula bar each time you click a step. This, however, involves understanding all the complexities of each piece of the code that underpins the data transformation process. I will provide a short overview of code modification in Chapter 14.

Tip   If you can force yourself to organize the process that you are writing with Power BI Desktop, then a little forethought and planning can reap major dividends. For instance, certain tasks, such as setting data types, can be carried out in a single operation. This means that you only have to look in one place for a similar set of data transformations. Not just that, but if you need to alter a data type for a column at a later stage, I suggest that you click the Changed Type step before you make any further alterations. This way, you extend the original step, rather than creating other steps—which can make the process more confusing and needlessly voluminous.

Renaming a Step

Power Query names steps using the name of the transformation that was applied. This means that if another similar step is applied later, Power Query uses the same name with a numeric increment. As this is not always comprehensible when reviewing a sequence of transformation steps, you may prefer to give more user-friendly names to individual steps. This is done as follows:

1. Select the query (or source table or worksheet, if you prefer). I will use the Clients query in this example.

2. Right-click the step that you want to rename, Changed Type, for instance.

3. Select Rename from the context menu.

4. Type in the new name. I will use **NewDataTypes**.

5. Press Enter.

The step is renamed, and the new name will appear in the Applied Steps list in the Query Settings pane. This way, you can ensure that when you come back to a data transformation process days, weeks, or months later, you are able to understand more intuitively the process that you defined, as well as why you shaped the data like you did.

Tip   You can—as you might expect—also simply double-click a step to rename it.

Deleting a Step or a Series of Steps

Deleting a step is all too easy, but doing so can have serious consequences. This is because a data ingestion and transformation process is often an extremely tightly coupled series of events, where each event depends intimately on the preceding one. So deleting a step can make every subsequent step fail. Knowing which events you can delete without drastic consequences will depend on the types of process that you are developing as well as your experience with Power Query. In any case, this is what you should do if you need to delete a step:

1. Place the pointer over the process step that you want to delete, or select the step.

2. Click the cross (×) icon that appears to the left of the step name.

3. Select Delete. The Delete Step dialog *might* appear, as shown in Figure 13-1.

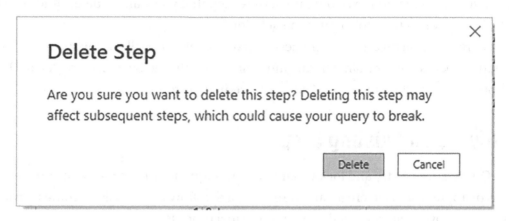

Figure 13-1. *The Delete Step dialog*

4. Confirm by clicking the Delete button. The step is deleted.

If—and it is highly possible—deleting this step causes issues for the rest of the process, you will see that the data table is replaced by an error message. This message will vary depending on the type of error that Power Query has encountered.

When describing this technique, I was careful to state that you *might* see the Delete Step dialog. If you are deleting the final step in a sequence of steps, then you will probably not see it, since there should not be any potentially horrendous consequences; at worst, you will have to re-create the step. If you are deleting a step in the middle of a process, then you might want to think seriously about doing so before you cause a potentially vast number of problems. Consequently, you are asked to confirm the deletion in these cases.

Note   If you realize at this point that you have just destroyed hours of work, then (after drawing a deep breath) click the File menu in the Power Query window (the downward-facing triangle at the top left) and select Close, and then close Power BI Desktop without saving. Don't count on using an undo function as you can in other desktop applications. To lower your blood pressure, you may prefer to save a copy of a file containing an intricate data transformation process *before* deleting any steps. You can also make copies of the entire data transformation process as "M" code—as you will learn in the next chapter.

An alternative technique is to right-click the step that you want to delete and select Delete. You may still have to confirm the deletion.

If you realize that an error in a process step has invalidated all your work up until the end of the process, rather than deleting multiple elements one by one, click Delete Until End from the context menu at step 2 in the preceding exercise.

Modifying an Existing Step

Power Query does not try and lock you into a rigid sequence of events when you create a series of applied steps to create and transform a dataflow. This really becomes obvious when you discover that you need to alter a step in a process.

Suppose, for instance, that you discover that you have loaded a wrong Excel worksheet when you selected the initial data from Excel. You do not want to repeat the process when you can simply substitute one worksheet name for another.

1. Select the query that you want to modify (Clients in this example).

2. Click the step to modify (in this case, it will be Navigation).

3. Click the gear (or cog) icon to the right of the step name.
   The appropriate dialog will appear. In this case, it will be the
   Navigation dialog that you can see in Figure 13-2.

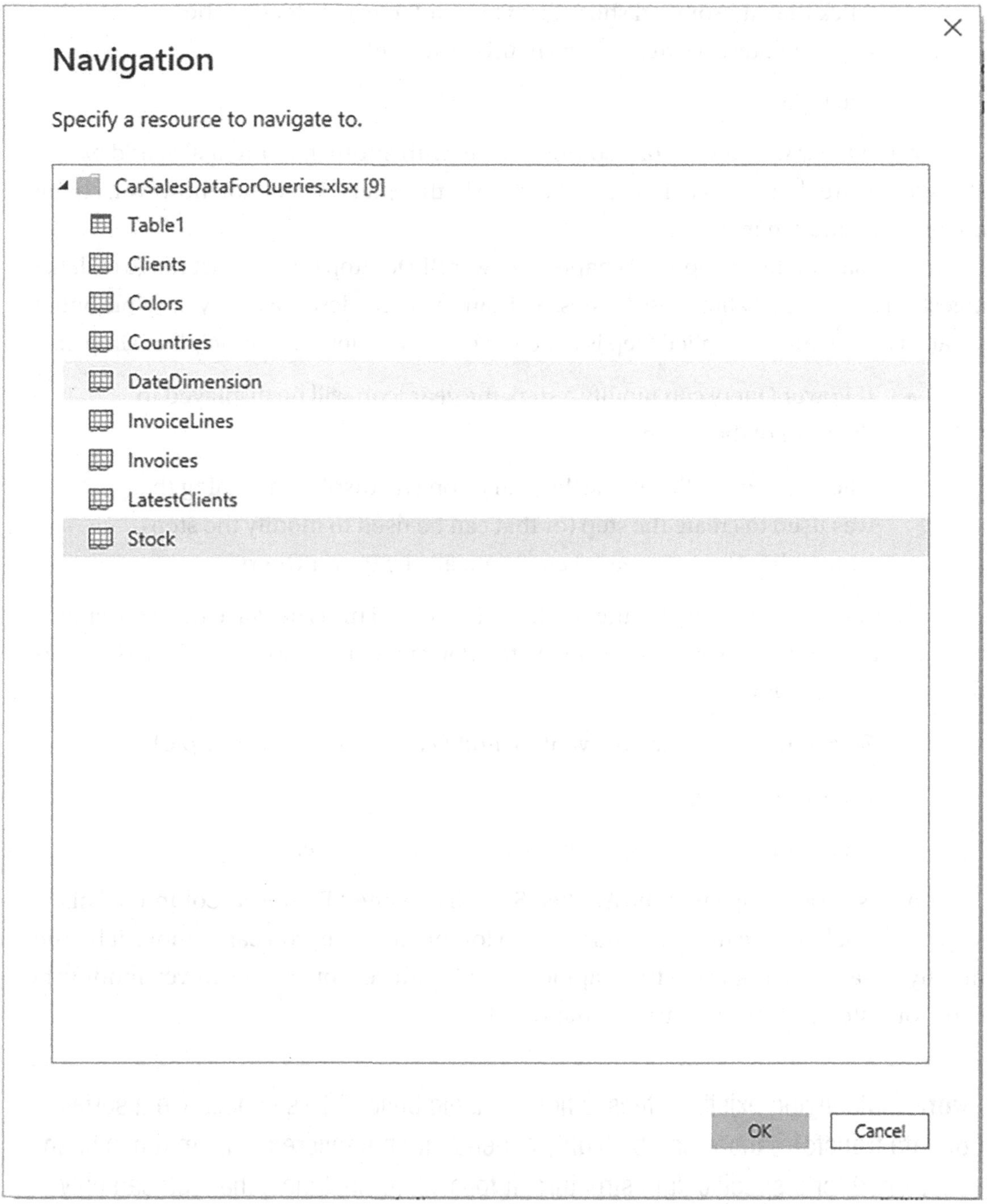

Figure 13-2. *The Navigation dialog displayed for step modification*

4. Click the table or worksheet that you want to use instead of the current dataset (LatestClients in this example).

5. Click OK.

Power Query will replace one source dataset with another. It might also add extra steps to ensure that the data is adapted for use in the query. It will not, however, change the original query name.

As you saw in the previous 12 chapters, Power BI Desktop offers a vast range of data ingestion and modification possibilities. So I cannot, here, describe every possible option as far as modifying an Applied Step is concerned. Nonetheless, the principle is simple:

- If Power Query can modify a step, the gear icon will be displayed to the right of the step name.

- Clicking the modification (the gear) icon will display the dialog that was used to create the step (or that can be used to modify the step even if the step was created automatically by Power Query).

Certain steps do not display the modification icon. This is because the step cannot be modified, only removed (at least, using the Power Query interface). As an example of this, add the following step:

1. Select the query that you want to modify (Clients in this example).

2. Click the last step.

3. Right-click the Address2 column and select Remove.

A new step will appear in the Applied Steps list, named Removed Columns. This step does not have the modification icon. So for the moment, you can remove it but *not* modify it—at least, not using the graphical user interface. You can, however, modify the code for a step as you will learn in Chapter 14.

Note   Modifying existing steps is not a "magic bullet." This is because a series of data transformations can be highly dependent on a tailored logic that has been developed for a specific data structure. It follows, for instance, that you can only replace a data source with another one that has a virtually identical structure. However, modifying a step can avoid your having to rewrite an entire dataflow sequence in many cases.

Adding a Step

You can add a step anywhere in the sequence. All you have to do is click the step that *precedes* the new step that you want to insert *before* clicking the icon in any of the ribbons that corresponds to the new step. As is the case when you delete a step, Power Query will display an alert warning you that this action *could* cause problems with the process from this new step on.

Altering Process Step Sequencing

It is possible—technically—to resequence steps in a process. However, in my experience, this is not always practical, since changing the order of steps in a process can cause as much damage as deleting a step. Nonetheless, you can always try it like this:

1. Right-click the step that you want to resequence.

2. Select Move Before or Move After from the context menu—or drag the step up or down if you prefer.

I remain pessimistic that this can work miracles, but it is good to know that it is there.

Tip   Remember that before tweaking the order in which the process is applied, clicking any process step causes the table in the Power BI Desktop window to display the snapshot of data up to the step that's been selected to show you the state of the data up to and including the selected step. This is a very clear visual guide to the process and how the data loading and transformation process is carried out.

An Approach to Sequencing

Given the array of available data transformation options, you may well be wondering how best to approach a new data load and transformation project using Power Query. I realize that all projects are different, but as a rough and ready guide, I suggest attempting to order your project like this:

1. Load the data into Power Query.

2. Promote or add comprehensible column headers. For example, you really do not want to be looking at step 47 of a process and wondering what Column29 is, when it could read (for instance) ClientName. Admittedly, Power Query often does this for you automatically–but it is always worth ensuring that columns have comprehensible names *earlier rather than later* in a data ingestion process.

3. Remove any columns that you do not need. The smaller the dataset, the faster the processing. What is more, you will find it easier to concentrate on, and understand, the data if you are only looking at information that you really need. Any columns that have been removed can be returned to the dataset simply by deleting or editing the step that removed them.

4. Set the appropriate data types for every column in the table. Correct data types are fundamental for many transformation steps and are essential for filtering, so it's best to get them sorted out *early* on even if Power Query often handles this automatically.

5. Filter out any records that you do not need. Once again, the smaller the dataset, the faster the processing. This includes deduplication.

6. Parse any complex JSON or XML elements.

7. Join or append any queries that can be assembled.

8. Carry out any necessary data cleansing.

9. Carry out any necessary transforms.

10. Carry out any necessary column splits or adding custom columns.

11. Add any derived columns.

12. Add any calculations or logical transformations of data.

13. Handle any error records that the data transformation process has thrown up.

Once again, I must stress that this is not a definitive guide. I hope, however, that it will help you to see "the wood for the trees" when you are creating data load and transformation processes using Power Query.

Error Records

Some data transformation operations will cause errors. This can be a fact of life when mashing up source data. For instance, you could have a few rows in a large dataset where a date column contains a few records that are texts or numbers. If you convert the column to a date data type, then any values that cannot be converted will appear as error values.

Removing Errors

Assuming that you do not need records that Power Query has flagged as containing an error, you can remove all such records in a single operation:

1. Click inside the column containing errors; or if you want to remove errors from several columns at once, Ctrl-click the titles of the columns that contain the errors.

2. In the Home ribbon, click Remove Errors in the Remove Rows pop-up menu. Any records with errors flagged in the selected columns are deleted. Removed Errors is added to the Applied Steps list.

You have to be very careful here not to remove valid data. Only you can judge, once you have taken a look at the data, if an error in a column means that the data can be discarded safely. In all other cases, you would be best advised to look at cleansing the data or simply leaving records that contain errors in place. The range and variety of potential errors are as vast as the data itself.

Managing Queries

Once you have used Power Query for any length of time, you will probably become addicted to creating more and deeper analyses based on wider-ranging data sources. Inevitably, this will mean learning to manage the data sources that feed into your data models efficiently and productively.

Fortunately, Power Query comes replete with a small arsenal of query management tools to help you. These include

- Organizing queries

- Grouping queries

- Duplicating queries

- Referencing queries

- Documenting queries

- Adding a column as a new query

- Enabling data load

- Enabling report refresh

Let's take a look at these functions, one by one.

Organizing Queries

When you have a dozen or more queries that you are using in Power Query, you may want to exercise some control over how they are organized. To begin with, you can modify the order in which queries appear in the Queries pane on the left of the Power Query window. This lets you override the default order, which is that the most recently added data source appears at the bottom of the list.

Do the following to change the position of a query in the list:

1. Right-click the query that you want to move.

2. Select Move Up (or Move Down) from the context menu—or just drag the query up or down.

You have to carry out the Move operation a number of times to move a query up or down a number of places if you are not dragging and dropping the query.

Grouping Queries

You can also create custom groups to better organize the queries that you are using in a Power BI Desktop file. This will not have any effect on how the queries work. Grouping queries is simply an organizational technique, and it will *not* change in any way the data tables that you see in report mode in Power BI Desktop.

Creating a New Group

The following explains how to create a new group:

1. Right-click the query that you want to add to a new group. I will use the Colors query. The context menu will appear as shown in Figure 13-3.

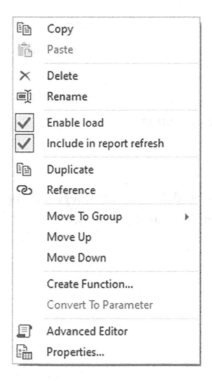

Figure 13-3. *Query context menu*

2. Select Move To Group ➤ New Group from the context menu. The New Group dialog will appear.

3. Enter a name for the group and (optionally) a description. I will name the group **Reference Data**. The dialog will look something like Figure 13-4.

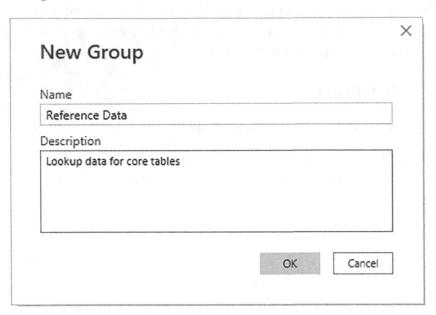

Figure 13-4. *The New Group dialog*

4. Click OK.

The new group is created, and the selected query will appear in the group. The Queries pane will look something like Figure 13-5.

Figure 13-5. *The Queries pane with a new group added*

Note   By default, all remaining queries that have not been added to a specific group become part of a group named Other Queries. The Other Queries group cannot be renamed or deleted. By default, all new queries will be added to this group.

Renaming Groups

You can rename any groups that you have added.

1. Right-click the query that you want to rename.

2. Select Rename from the context menu.

3. Edit or replace the name.

4. Press Enter.

Tip   You can also double-click the query name to rename it.

Adding a Query to a Group

To move a query from its current group to another group, you can carry out the following steps:

1. Right-click the query that you want to add to another existing group.

2. Select Move To Group ➤ *Destination Group Name* from the context menu.

The selected query is moved to the chosen group.

Duplicating Queries

If you have done a lot of work transforming data, you could well want to keep a copy of the original query before trying out any potentially risky alterations to your work. Fortunately, this is extremely simple.

1. Right-click the query that you want to copy.

2. Select Duplicate from the context menu.

The query is copied, and the duplicate appears in the list of queries inside the same group as the source query. It has the same name as the original query, with a number in parentheses appended. You can always rename it in the Query Settings pane or in the Queries pane on the left of the Power Query window.

Note   You can copy and paste queries if you prefer. The advantage of this technique is that you can choose the destination group for the copied query simply by clicking the folder icon for the required group *before* pasting the copy of the query.

Referencing Queries

If you are building a complex ETL (Extract, Transform, Load) routine, you might conceivably organize your work in stages to better manage the process. To help you with this, Power Query allows you to *use the output from one query as the source for another query*. This enables you to break down different parts of the process (e.g., structure, filters, then cleansing) into separate queries so that you can concentrate on different aspects of the transformation in different queries.

To use the output of one query as the source data for another, you need to *reference* a query. The following explains how to do it:

1. Right-click the query that you want to use as the source data for a new query.

2. Select Reference from the context menu. A new query is created in the list of queries in the Queries pane.

3. Right-click the new query, select Rename, and give it a meaningful name.

Unless you rename the query, the new query has the same name as the original query, with a number in parentheses appended. If you click the new query, you see exactly the same data in the referenced query as you can see if you click the final step in the source query.

From now on, any modifications that you make in the referenced (source) query produce an effect on the data that is used as the source for the second query.

In practice, I suspect, you will not want to use two copies of the same query to create reports. Indeed, if a query is being used as an "intermediate" query, the data that it contains might not even be fully usable. So you could want to *make the intermediate query unavailable to reports and dashboards* in Power BI Desktop. To do this:

1. Right-click the original (source) query.

2. Deselect Enable Load from the context menu. The check mark to the left of the menu item will disappear, and the query name will appear in italics. You can see this in Figure 13-6.

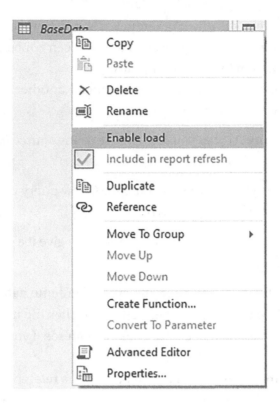

Figure 13-6. *Preventing query load into Power BI Desktop*

The source query will no longer appear in Report View and so cannot be used in visuals. It is worth noting that any queries that are "intermediate" queries (i.e., queries that you use to modify data but that do not show in Data View or Report View) are in italics.

You may be wondering why you would want to create "intermediate" queries. Some ideas are

- You want to isolate complex data transformations into more manageable subsets. You may, for instance, want one intermediate query that transforms the data while a subsequent query cleanses the data.

- You could want to apply a common set of initial transformations that then feed into two separate data preparation paths—a detailed view of the data and an aggregated view.

Note   Ensure that any existing reports do not use a query that you subsequently make unavailable in this way, or you will end up with broken visuals in your reports.

Documenting Queries

In a complex ETL process, it is easy to get confused—or simply forget—which query does what. This can be important not only for yourself but also for people inheriting the maintenance responsibility of a data ingestion process that you have built Consequently, I always advise documenting queries by adding a meaningful description.

1.   Right-click the query that you want to annotate.

2.   Select Properties from the context menu. The Query Properties dialog will appear.

3.   Add a description. The result could be like the dialog shown in Figure 13-7.

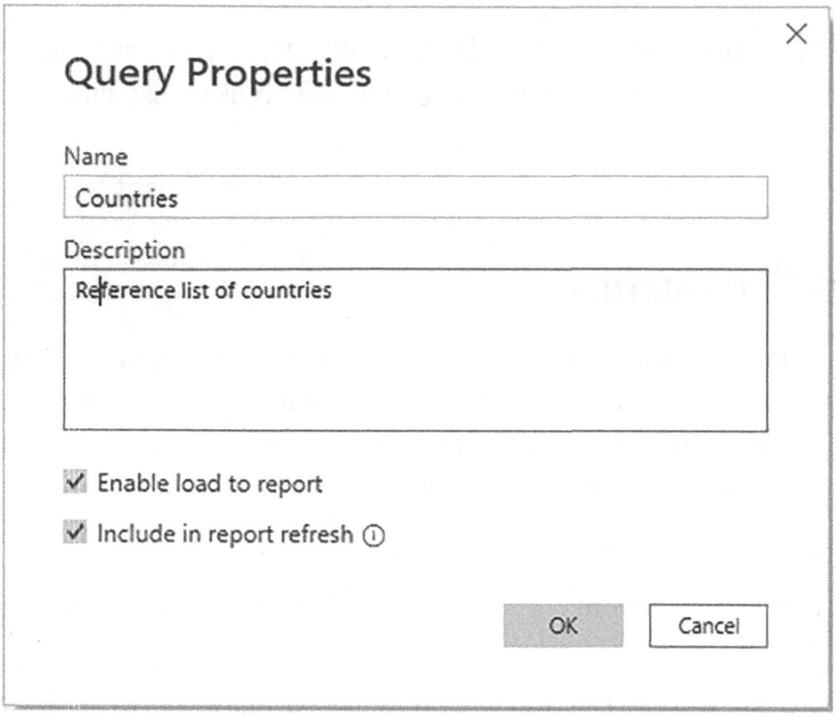

Figure 13-7. *Adding a description to a query*

4. Click OK.

The description that you added is now visible as a tooltip if you hover the cursor over the query name in the list of queries in Power Query.

Adding a Column As a New Query

There are occasions when you might want to extract a column of data and use it as a separate query. It could be that you need the data that it contains as reference data for another query, for example. The following steps explain how you can do this:

1. In the Queries list on the left, select the query containing the column that you want to isolate as a new query. I will use the Countries query.

2. Right-click the title of the column containing the data that you want to isolate. I will use CountryName.

3. Select Add as New Query from the context menu. A new query is created. It is named after the original query and the source column. You can see this in Figure 13-8.

	List
1	United Kingdom
2	France
3	USA
4	Germany
5	Spain
6	Switzerland

Figure 13-8. *A list created from a column*

4. In the List Tools Transform ribbon (generally at the right of the classic menus), click To Table. The To Table dialog will appear, as you can see in Figure 13-9.

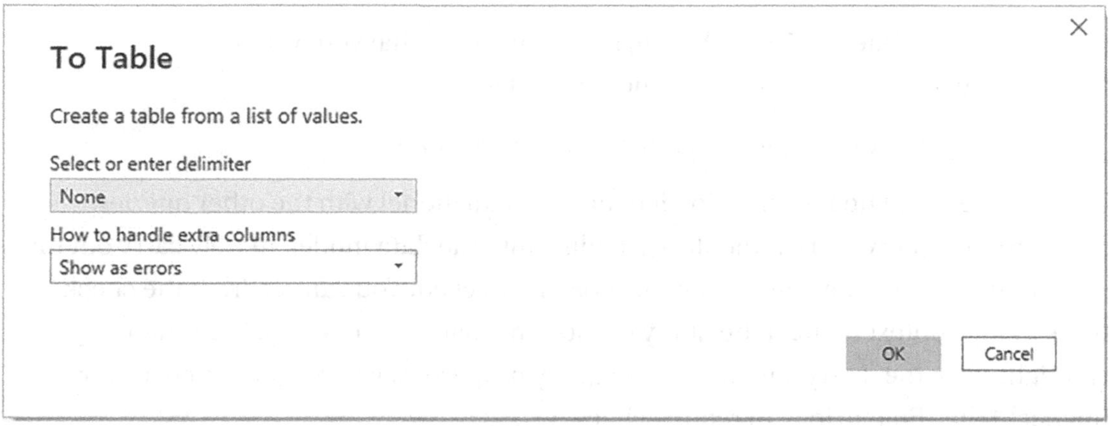

To Table

Create a table from a list of values.

Select or enter delimiter

None	▾

How to handle extra columns

Show as errors	▾

OK   Cancel

Figure 13-9. *The To Table dialog*

5. Click OK. The new query will become a table of data and will have the name of the column that you selected. Note that the column name itself has now disappeared, and it's just called "Column1."

6.   Rename the query, if you judge this necessary.

You can now use this query in your data model and as part of a linked set of query processes.

Note   A query created in this way is *completely disconnected* from the source query from where the data was taken. Put another way, any refresh of the source data will have *no* effect on the new query that you created from a column.

Enabling Data Load

You may have gained the impression that any query that you create will always become part of the Power BI Desktop data model. This is emphatically *not* the case as I mentioned briefly in the section on reference queries. You can create queries that are in effect only "staging" queries that are part of a more complex sequence of transformations or queries that contain only lookup data that is added to another table but not needed in the data model, for instance. In cases like these, you certainly do not want these tables adding clutter to the data model.

To prevent a query being added to the data model, do the following:

1.   In the Queries list on the left, select the query that you want to keep in Power Query—but not in the data model.

2.   Right-click the query and uncheck Enable Load.

The query will no longer be loaded into the data model with the other queries.

To reset a query as a candidate for loading into the data model, merely carry out the same operation and ensure that Enable Load is checked. You can see from the check mark in the context menu if the query is due to be loaded or not. As an alternative, you can right-click the query and display the query properties, where you can check (or uncheck) the "Enable load to report" check box.

Note   You can also set this property in the Query Properties dialog that you saw in Figure 13-7.

Enabling Report Refresh

By default, all queries can be refreshed. This lets you gather the very latest data from the source into Power Query and then into the data model.

There could be times when you do not want to refresh a query. Perhaps the data source is unobtainable, or is slow, or you want to return later for the latest data. Power BI Desktop Power Query lets you set the refresh option for each query like this:

1. In the Queries list on the left, right-click the query whose refresh status you want to modify.

2. Uncheck "Include in report refresh." You can see this in Figure 13-10.

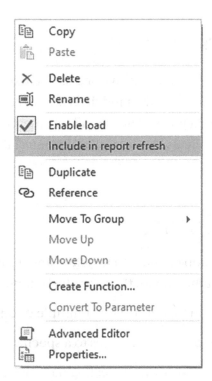

Figure 13-10. *Preventing report refresh*

Note   Only queries that are enabled for load can have the refresh property modified.

Pending Changes

When you are dealing with data sources and switch from Power Query to the Power BI Desktop view, normally, you then want to load the full data from the source into the data model.

The downside to this approach is that a huge set of source data can take a long time to load, or reload, when you move back to creating and modifying visualizations. This is why Power Query will let you select Close from pop-up options available in the Close & Apply button in the Home ribbon. Doing this will return you instantly to the Data View but will not apply any changes that you have made. As a reminder, you will see an alert like the one shown in Figure 13-11 at the top of the Data View.

Figure 13-11.   *The pending changes alert*

You can continue to work in Data View as long as you like. Click the Apply changes button when you have time to reload the modified source data into the data model. You need to be aware, however, that you will not see all the latest changes to the structure of the tables and fields in the Fields list—as well as the latest data, of course—until you apply the changes.

Parameterizing Queries

Parameters in Power BI Desktop enable you to define and apply specific criteria to certain aspects of queries. At their heart, they are a technique that enables you to

- Store a value that can then be used in multiple queries

- Restrict a selection of potential values to a specific list of options

There are currently three basic ways of creating parameters:

- A single value that you enter

- A selection of a value from a list of possible values that you enter manually

- A selection of a value from a list of possible values that you create using existing queries

It follows that using parameters is a two-step process:

- Create a parameter.

- Apply it to a query.

A parameter is really nothing more than a specialized type of query. As it is a query, you can

- Load it into the data model (although this is rarely required)

- Reference it from another query

The next three short subsections will explain how you can create parameters. I will then show you some of the ways that you can apply parameters in Power Query to filter or transform the data.

Creating a Simple Parameter

At its simplest, a parameter is a value that you store so that you can use it later to assist you in your data transformation. Here is how you can store a parameter containing a "True" value ready for use in filtering subsequent datasets:

1. Load the data contained in the Excel file
   `C:\PowerBiDesktopSamples\CarSalesOverview.xlsx` (using the worksheet that is also called CarSalesOverview) into Power BI Desktop.

2. In the Home ribbon, click Transform Data. Power Query will open.

3. In the Power Query Home ribbon, click the small triangle at the bottom of the Manage Parameters button, and then select New Parameter from the available menu options. The Parameters dialog will appear.

4. Enter **DealerParameter** as the parameter name and **Filter dealer types** as the description.

5. Choose True/False from the pop-up list of types.

6. Enter True as the Current Value from the pop-up list. The dialog will look like the one in Figure 13-12.

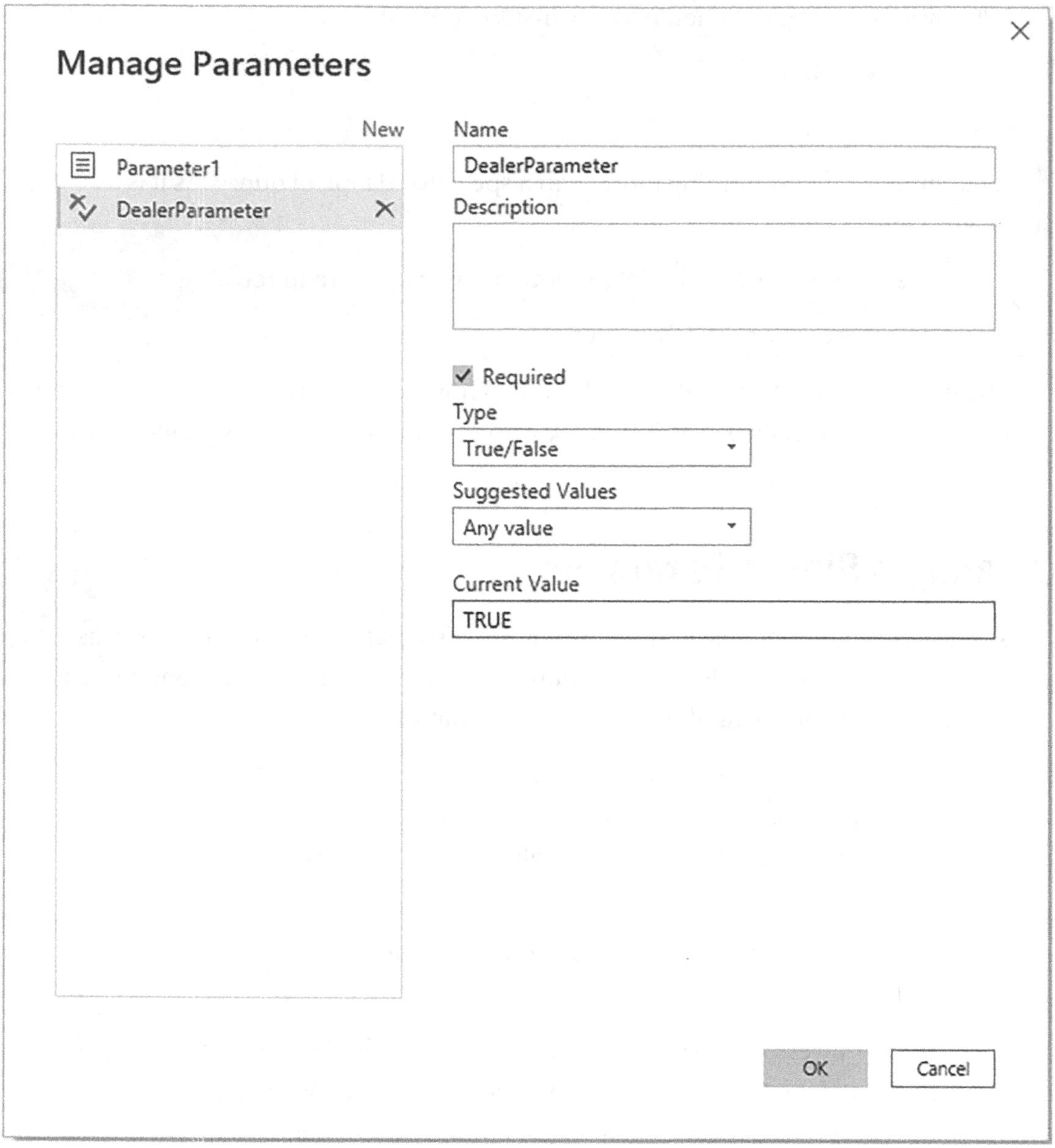

Figure 13-12.   *The Parameters dialog*

7.   Click OK. The new parameter will appear in the Queries list on the
     left. You can see this in Figure 13-13.

Figure 13-13. *A parameter in the Queries list*

For the moment, all you have done is create a parameter and store a value in it. You will see how to use this parameter in a few pages' time. As you can see, a parameter is stored as a type of query, and the default value is displayed after the query name in parentheses. Moreover, parameters always appear in italics in the Queries pane.

Creating a Set of Parameter Values

While a single parameter can always be useful, in reality, you are likely to need lists of potential parameters. This will allow you to choose a parameter value from a predefined list in certain circumstances. Here is an example of creating a parameter containing a subset of the available country names used in the sample data:

1. Using the Power BI Desktop file that you created in the previous section (the one based on the Excel file
    `C:\PowerBiDesktopSamples\CarSalesOverview.xlsx`),
    open Power Query unless Power Query is already open.

2. In the Power Query Home ribbon, click the small triangle at the bottom of the Manage Parameters button, and then select New Parameter from the available menu options. The Parameters dialog will be displayed.

3. Enter **CountriesParameter** as the parameter name.

4. Ensure that the Required check box is selected.

5. Choose Text from the pop-up list of types.

6. In the Suggested Values pop-up list, select List of Values.

7. Enter the following three values in the grid that has now appeared:

    a. France

413

      b. Germany

      c. Spain

8. Select France as the Default Value from the pop-up list.

9. Select France as the Current Value from the pop-up list. The dialog will look like the one shown in Figure 13-14.

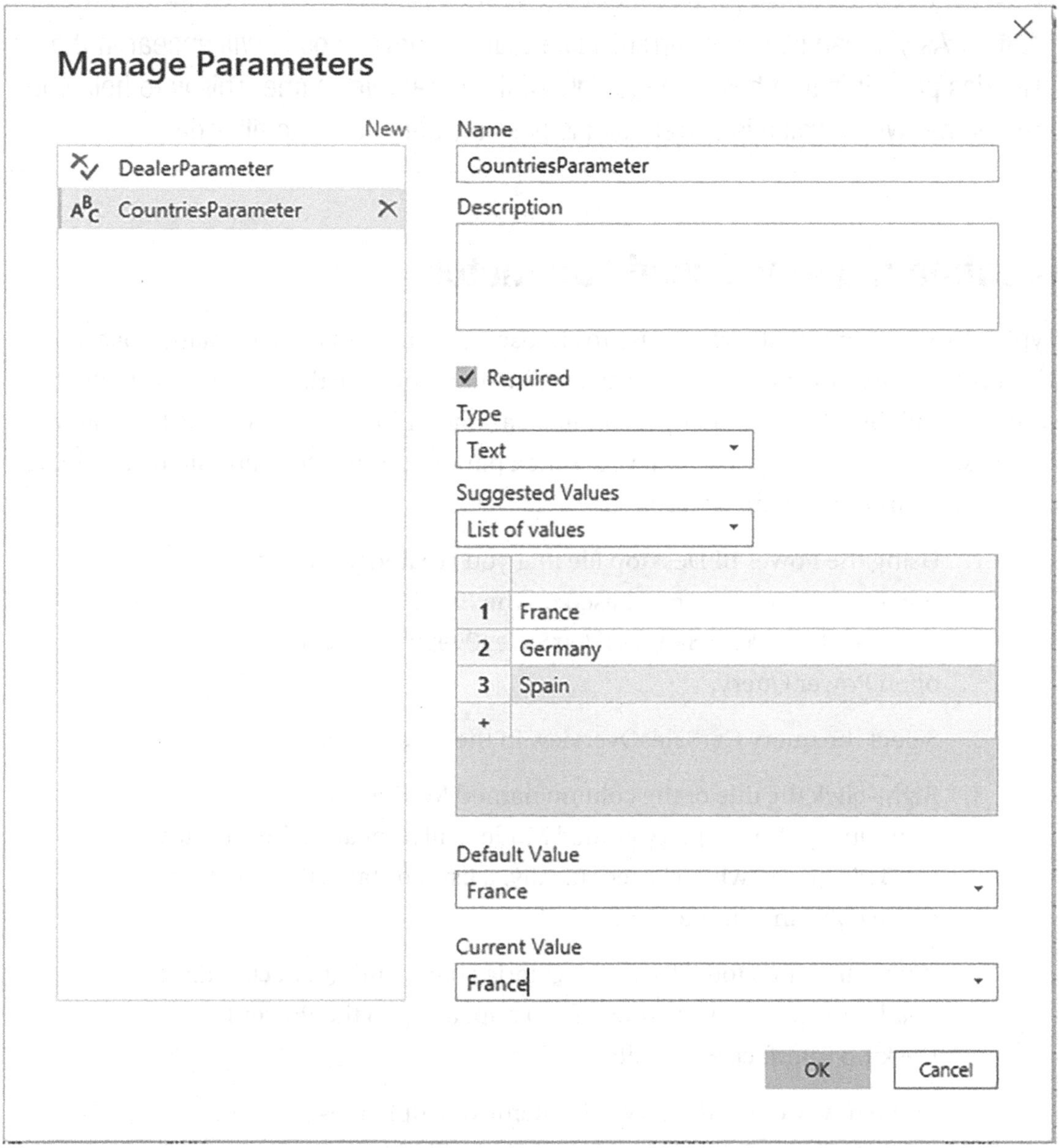

Figure 13-14. *The Parameters dialog*

10. Click OK. The new parameter will appear in the Queries list on
     the left.

Once again, all you have done is create the parameter. You will see how it can be
applied in a couple of pages' time.

Note   As you can see, any current value that you have chosen will appear in the Queries pane in parentheses to the right of the parameter name. This is to help you remember which value is current and is possibly being used to filter data.

Creating a Query-Based Parameter

Typing lists of values that you can use to choose a parameter is not only laborious, it is also potentially error-prone. So you can use the data from existing queries to create the series of available elements that you use in a parameter instead of typing lists of values. As an example of this, suppose that you want a parameter that contains all the available makes of car that the company sells:

1. Using the Power BI Desktop file that you created in the previous section (the one based on the Excel file C:\PowerBiDesktopSamples\CarSalesOverview.xlsx), open Power Query.

2. Select the query CarSalesOverview in the Queries list.

3. Right-click the title of the column named Model, and select Add as New Query. A new query named Model will appear in the Queries list. This query (which is, technically, a list) contains the contents of the column you selected.

4. Right-click the Model list in the Queries pane and uncheck Enable Load. This prevents the query from appearing in the Power BI Desktop interface as a table.

5. In the newly created query, click Remove Duplicates in the Transform ribbon. The List column will only display unique values.

6. Rename the newly created query **ModelList**.

7. In the Power Query Home ribbon, click the small triangle at the bottom of the Manage Parameters button, and then select New Parameter from the available menu options. The Parameters dialog will be displayed.

8. Enter **ModelsParameter** as the parameter name.

9. Ensure that the Required check box is selected.

10. Choose Text from the pop-up list of types.

11. In the Suggested Values pop-up list, select Query.

12. Select ModelList as the query containing a list of values to use from the pop-up list of available lists.

13. Enter DB9 as the Current Value. The Manage parameters dialog will look like Figure 13-15.

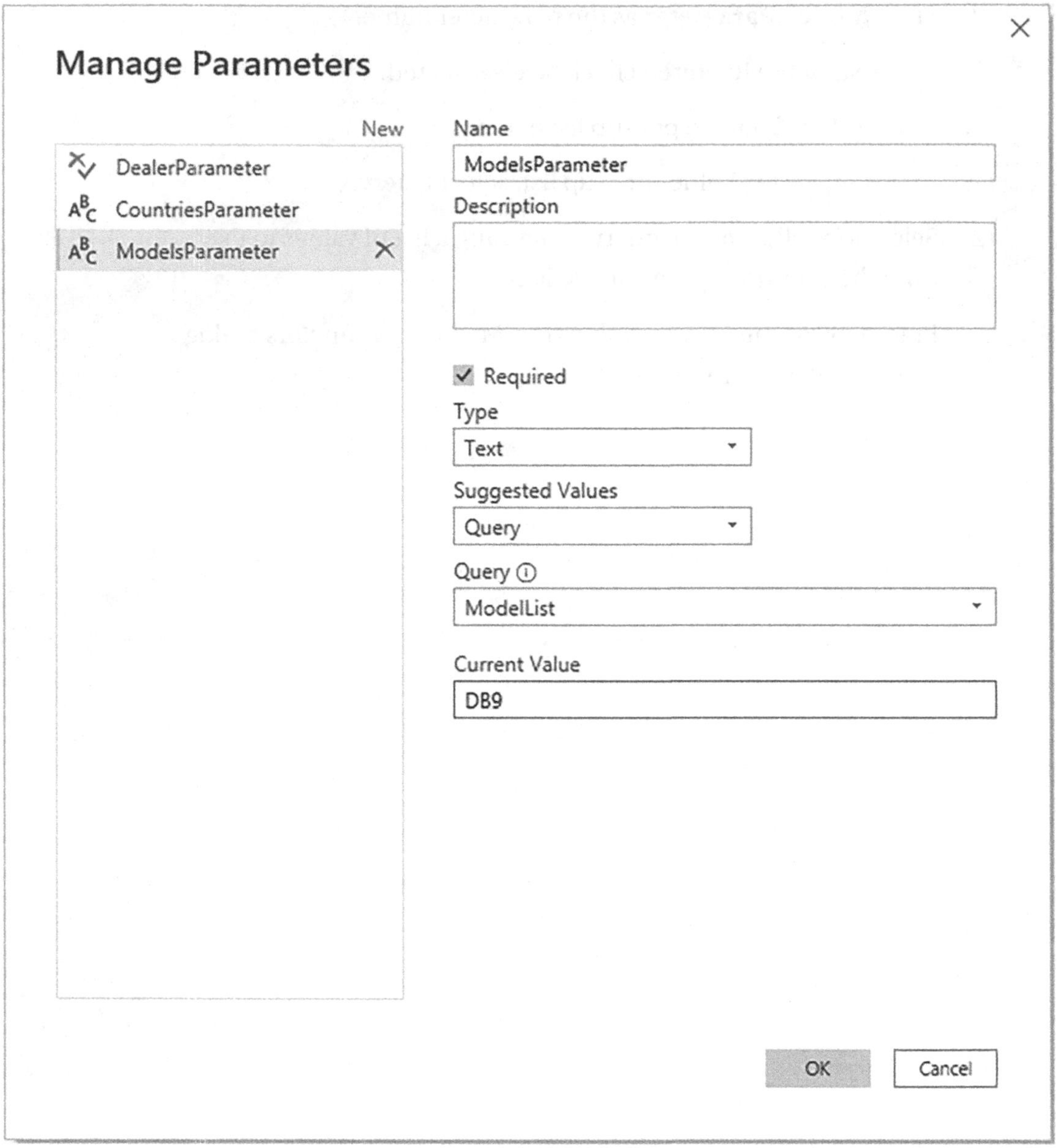

Figure 13-15.   *Using a list as parameter values*

14.   Click OK. The new parameter will appear in the Queries list on
the left.

You should now be able to see all three parameters that you have created in the
Queries pane, as shown in Figure 13-16.

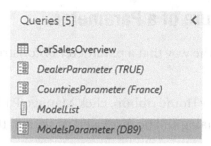

Figure 13-16. *Parameters in the Queries list*

Note   It is normal to prevent a list that is being used to provide a series of values for a parameter from being loaded into the data model (as you did in step 4). This is because this query (or list, if you prefer) is not required in the data model, and you do not want to confuse users by having it appear in the Fields list.

Once your parameters have been created, you can quit Power Query by clicking the Close & Apply button. You should see no change in the Power BI Desktop Report View. This is because parameters are used to shape a dataflow process, but *not* as the data in the data model.

Tip   It is also possible to create parameters "on the fly" (which is directly from inside a dialog that uses a parameter) when you want to use them. However, I find it better practice—and more practical—to prepare parameters beforehand. This forces you to think through the reasons for the parameter as well as the potential range of its use. It can also avoid your making errors when trying to do two different things at once.

Modifying a Parameter

Fortunately, parameters are not set in stone once they are created. You can easily modify

- The structure of a parameter

- The selected parameter element (the current value)

Modifying the Structure of a Parameter

Should you need to modify the way that a parameter is constructed, one way is to do the following:

1. In the Power Query Home ribbon, click Manage Parameters. The Parameters dialog will be displayed as seen in the previous sections.

2. In the left pane of the dialog, click the parameter that you want to modify. The parameter definition will appear on the right.

3. Carry out any required modifications.

4. Click OK.

Alternatively, you can do this:

1. Click the parameter in the Queries pane on the left of Power Query.

2. Click the Manage Parameter button. The Parameters dialog will appear.

3. Carry out any required modifications.

4. Click OK.

You can also, if you prefer, right-click a parameter in the Queries pane and select Manage from the pop-up menu to display the Parameters dialog.

Applying a Parameter When Filtering Records

Now that you have seen how parameters are created, it is time to see them in action. As a first example of applying a parameter, you will see how to use a parameter to filter a query:

1. In Power Query, create the three parameters described previously using the Excel file CarSalesOverview.xlsx as the data source.

2. Click the CarSalesOverview query in the Queries list. A dataset of car sales information will appear.

3. Click the pop-up menu (the down-facing triangle at the right of the column name) for the CountryName column.

4. Select Text Filters ➤ Equals. The Filter Rows dialog will appear.

5. Leave Equals as the first choice.

6. Click the second popup and select Parameter from the list. You can see this in Figure 13-17.

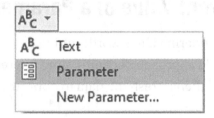

Figure 13-17. *Selecting a parameter for a filter*

7. Select CountriesParameter for the third popup. The dialog will look like the one shown in Figure 13-18.

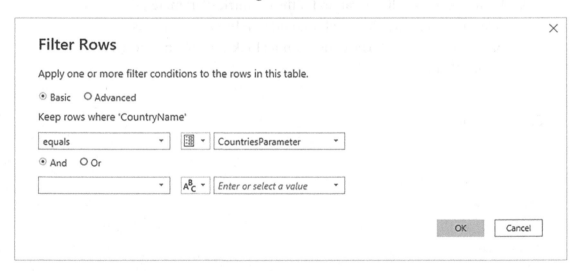

Figure 13-18. *Applying a parameter for a filter*

8. Click OK. The current parameter value (the country that you selected) will be applied, and the dataset will be filtered using the current parameter value.

Note   To remove a parameter from a filter, simply delete the relevant step in the Applied Steps list. Alternatively, you can edit the step and select Text instead of Parameter from the middle popup and then enter a "hard-coded" element to filter on.

Modifying the Current Value of a Parameter

You could be forgiven for wondering if it is worth setting up a parameter merely to filter a dataset. However, this whole approach becomes more interesting if you modify the current parameter value and then refresh the data to apply the new parameter. Here is an example of this:

1. In the Power Query Home ribbon, click the small triangle to display the menu for the Manage Parameters button.

2. Select Edit Parameters. The Enter Parameters dialog will appear.

3. From the pop-up list of values for the CountriesParameter, select one of the available values (and not the value that was previously selected). The dialog should look like the one shown in Figure 13-19.

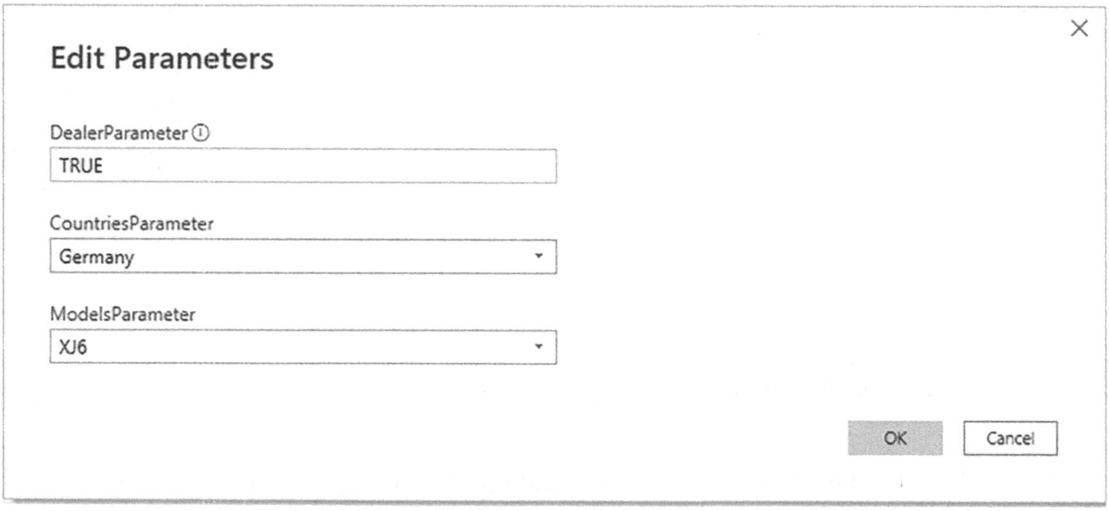

Figure 13-19. *Modifying the current value of a parameter*

4.   Click OK.

5.   In Power Query, the data will be refreshed and the new parameter values applied to the filters that use these parameters.

This approach becomes particularly useful if you have many combinations of filter values to test. In essence, you can apply a series of filters to several columns (or create complex filters) using several parameters and then test the results of different combinations of parameters on a dataset using the Enter Parameters dialog. This technique avoids having to alter multiple filters manually—and repeatedly. As an added bonus, you can restrict the user (or yourself) to specific lists of parameter choices by defining the lists of available parameter options. You can see this for the pop-up lists that appear when you select the CountriesParameter popup or the ModelsParameter popup.

Applying a Parameter in a Search and Replace

Another use for parameters is to apply them as either the search value or the replacement value in a search and replace operation. You can see this in the following example:

1.   In Power Query, create the CountriesParameter parameter as described previously.

2.   Click the CarSalesOverview query in the Queries pane.

3.   Click inside the CountryName column.

4.   In the Transform ribbon, click Replace Values (or right-click and select Replace values). The Replace Values dialog will be displayed.

5.   Click the pop-up list to the left under Value To Find, and select Parameter.

6.   Choose CountriesParameter as the parameter to apply.

7.   Enter Luxembourg (for instance) as the replacement value. The dialog will look like the one in Figure 13-20.

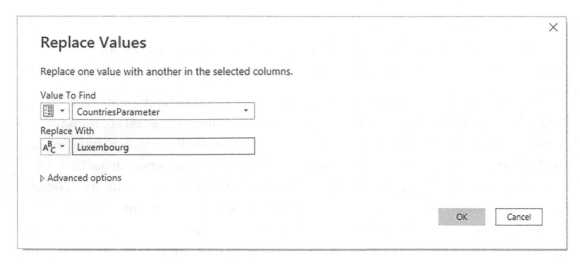

Figure 13-20. *Using a parameter in search and replace*

8. Click OK. The parameter's current value will be replaced in the
    dataset for the selected column.

Always Allow Parameters

The "Always Allow Parameters" check box is only relevant when you've not yet defined
any parameters in Power Query. If you have not yet defined any parameters and this
setting is turned off, you must create a parameter through the "normal" procedure before
you can use a parameter for any data source/transformation.

However, if you have not yet defined any parameters and this setting is turned on,
you can go to perform your transformation/connect to your data source as you would
normally and then create a parameter "on the fly" using the process that you saw
previously.

So the Always Allow Parameters setting is essentially just allowing you to create
parameters in any place you're allowed to use parameters, for a report that doesn't yet
have any parameters defined.

Applying a Parameter to a Data Source

In some corporate environments, there are many database servers that are available, and
possibly even more databases. You may find it difficult to remember all of these—and so
may the users that you are preparing Power BI Desktop reports for.

One solution that can make a corporate environment easier to navigate is to prepare parameters that contain the lists of available servers and databases. These parameters can then be used—and updated—to guide users in their choice of SQL Server, Oracle, or other database data sources.

To see this in action, you will first have to prepare two parameters:

- A list of servers

- A list of databases

You can then see how to use these parameters to connect to data sources. Of course, you will have to replace the example server and database names that I use here in steps 3 and 4 with names from your own environment.

1. Open a new Power BI Desktop file and close the splash screen.

2. Click Transform Data to open Power Query.

3. Create a new parameter (using the techniques defined previously) using the following elements:

   a. *Name*: Servers

   b. *Type*: Text

   c. *Suggested Values*: List of Values

   d. *Values in the list*: ADAM03 and ADAM03\SQLServer2019 (or your database server)

   e. *Default Value*: ADAM03\SQLServer2019 (or your database server)

   f. *Current Value*: ADAM03\SQLServer2019 (or your database server)

4. Create a new parameter using the following elements:

   a. *Name*: Databases

   b. *Type*: Text

   c. *Suggested Values*: List of Values

   d. *Values in the list*: PrestigeCars and PrestigeCarsDataWarehouse

   e. *Default Value*: PrestigeCars

   f. *Current Value*: PrestigeCars

5.   The two parameters will look like those displayed in Figure 13-21.

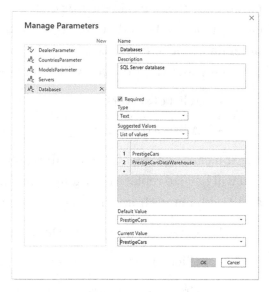

Server Parameter   Database Parameter

Figure 13-21.   *Defining parameters for server and database*

6.   Click Close & Apply to close Power Query.

7.   In the Home ribbon of the Power BI Desktop Report screen, click
     the SQL Server button.

8.   On the Server line, click the popup for the server and choose
     Parameters. Select the Servers parameter.

9.   On the Database line, click the popup for the database and choose
     Parameters. Select the Databases parameter. The SQL Server
     dialog will look like the one shown in Figure 13-22.

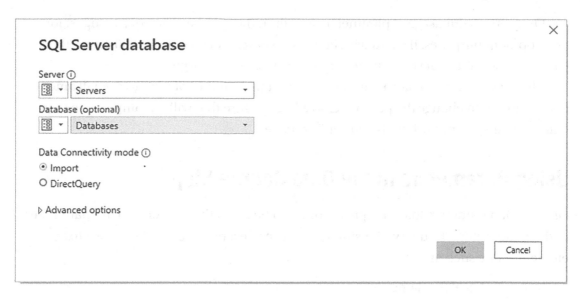

Figure 13-22. *Using a parameter to select the server and database*

10. Choose the data connectivity mode and any advanced options that you want to set.

11. Click OK. The server connection process and dialogs will appear, and you will then see the Navigator dialog displaying the tables and views for the current server and database values in the two parameters.

Note   Preparing parameters for data connection is particularly useful if you then save the file as a template that can be used as the basis for multiple different Power BI Desktop report files. You will learn how to create templates in a few pages.

Other Uses for Parameters

These examples only cover a few of the cases where parameters can be applied in Power BI Desktop. Indeed, the range of circumstances where a parameter can be applied is increasing with each release of the product. So look out for all the dialogs that give you the option of using a parameter!

One very useful usage of parameters is to limit the number of rows you ingest for development purposes. If you have a large(ish) source (let's say 1,000,000 rows), then you may not want to develop the report with a dataset that large.

This way, you can create a parameter to limit the number of rows to a small amount of data and then change the parameter to a larger figure that will encompass the entire dataset when you publish to the Power BI service.

Using Parameters in the Data Source Step

One use of parameters that can quickly prove to be a real time-saver is to use parameters in the Source step of a query. Put simply, you can use a parameter instead of a fixed element name such as

- An Excel file name

- A file path

- A database or data warehouse server

- A database

It can be particularly useful to use parameters to define connections (i.e., server and database references), as this

- Provides a central reference point for connection information

- Avoids you having to type connection details for similar queries from the same server—and minimizes the risk of introducing typos

- Makes it easier to switch between development, test, and production servers

To illustrate this, and assuming that you have created the parameters "Servers" and "Databases" from the previous section, try the following:

1. Create a connection to a SQL Server database (as described in Chapter 3).

2. Click the Transform Data button in the Home ribbon.

3. In Power Query, select the query created by the database connection.

4. Click the first of the Applied Steps on the right. This step should be named "Source."

5. In the formula bar, replace the code that looks something like this:

    ```
    = Sql.Database("ADAMO3\SQL2019", "PrestigeCars")
    ```

    with this:

    ```
    = Sql.Database(Servers, Databases)
    ```

6. Confirm your modifications by clicking the check box in the formula bar—or by pressing Enter. You will almost certainly have to confirm your database credentials.

Note   You need to be aware that hard-coded server and database names must be contained in double quotes, whereas parameters must *not* be enclosed in quotes. Also remember that the M language used in the formula bar is case-sensitive. So you need to enter parameter names *exactly* as they were created.

Applying a Parameter to a SQL Query

If you are using a relational database, such as Oracle or SQL Server, as a data source (and if you are reasonably up to speed with the flavor of SQL that the source database uses), you can query a database using SQL and then apply Power BI Desktop parameters to the source query.

Let's see this in action:

1. Open a new Power BI Desktop file, open Power Query, and create the parameter named CountriesParameter that you saw a few pages ago.

2. Click Close & Apply to close Power Query.

3. In the Power BI Desktop Report screen, click SQL Server database.

4. Enter the server and database that you are using.

5. Click Advanced options and enter the following SQL statement:

```
SELECT  *
FROM    CarSalesData.Data.CarSalesData
WHERE   CountryName = 'Germany'
```

6. Click OK and confirm any dialogs about data access and permissions.

7. Click Edit to connect to the data and open Power Query.

8. There should only be one Applied Step for the data connection. Expand the formula bar and tweak the formula so that it looks like this:

```
= Sql.Database("ADAMO3\SQLSERVER2019", "CarSalesData",
[Query="SELECT *#(lf)FROM    CarSalesData.Data.
CarSalesData#(lf)WHERE   CountryName =
'"& CountriesParameter &"'"])
```

9. Click the tick icon in the formula bar to confirm your changes (or press Enter). The data will change to display the data for France (the current parameter value) rather than Germany (the initial value in the SQL). You may need to give Permission to run the native database query.

You can now alter the parameter value and refresh the data. This will place the current parameter inside the SQL WHERE clause and only get the data for the current parameter.

In case this seems a little succinct, let's look at the code used by Power BI Desktop *before* you made the change in step 8. The M language read:

```
= Sql.Database("ADAMO3\SQLSERVER2016", "CarSAlesDAta",
[Query="SELECT *#(lf)FROM    CarSalesData.Data.CarSalesData#(lf)
WHERE   CountryName = 'Germany'#(lf)"])
```

The change was to replace

France

with

"& CountriesParameter &"

What you did was to replace the hard-coded criterion "France" with the parameter reference. Indeed, much as you would in Excel, you added double quotes and ampersands to the formula to allow the code to include an extraneous element.

This was an extremely simple example, but I hope that it opens the door to some fairly advanced use of parameters in database connections.

Note   Updating data once a parameter has changed might require accepting data changes and new permissions.

Query Icons

As you could see previously in Figure 13-16, there are three query icons. These are explained in Table 13-1.

Table 13-1.   *Query Icons*

Icon	Query Type	Description
▦	Query	The icon for a standard query
▯	List	The icon for a list
▤	Parameter	The icon for a parameter

Power BI Templates with Parameters

If you have created a Power BI model that contains powerful data transformation routines or simply a set of data connection options—like the ones that you saw in the previous pages—you could want to use any or all of these as the common basis for multiple reports.

The solution to this is extremely simple. You just save a Power BI Desktop file containing all the code, ETL, and other resources that you will need in future reports as a template. You can then open the template and a new Power BI Desktop file will open using all the template's queries, parameters, and code. You then save the new file perfectly normally as a "standard" Power BI Desktop file.

To save a file as a template:

1. Create the model file containing the queries, parameters, and all other elements that make it the basis for future reports.

2. In the Home ribbon, click File ➤ Save As.

3. In the Save As dialog, select Power BI Template File (∗.pbit) as the Save As type.

4. Enter a file name and click OK.

You can now open the template file just as you would any other Power BI Desktop file. Power BI Desktop will open a copy of the template, leaving the template itself untouched. If the template file contains parameters, the user will be prompted to enter parameter values when the template is opened.

Conclusion

In this chapter, you saw how to manage and extend the contents of the queries that you can create using Power BI Desktop. Specifically, you saw how to modify individual steps in a data load and transformation process. This ranged from renaming steps to changing the order of steps in a process—or even altering the specification of what a step actually does.

Then you saw how to manage whole queries. You learned how to rename and group queries as well as how to chain queries so that the output from one query became the source of data for another query.

Finally, you learned how to add parameters to queries and how to interact with queries in a controlled fashion. This lets you make queries—and so the entire ETL process—more flexible and interactive.

CHAPTER 14

The M Language

Data ingestion and modification are not only interface driven in Power BI Desktop. In fact, the entire process is underpinned and powered by a highly specific programming language. Called "M," this language underlies everything that you have learned to do in the last 13 chapters.

Most users—most of the time—are unlikely to need to use the M language directly at all. This is because the Power Query Editor interface that you have learned so much about thus far in this book is both comprehensive and extremely intuitive. Yet there may be times when you will need to

- Add some additional functionality that is not immediately accessible through the graphical interface

- Add programming logic such as generating sequences of dates or numbers

- Create or manipulate your own lists, records, or tables programmatically

- Create your own built-in functions to extend or enhance those that are built into the M language

- Use the Advanced Editor to modify code

- Add comments to your data ingestion processes

Before introducing you to these concepts, I need to add a few caveats:

- The "M" language that underpins Power BI Desktop queries is not for the faint of heart. The language can seem abstruse at first sight.

- The documentation is extremely technical and not wildly comprehensible for the uninitiated.

- The learning curve can be steep, even for experienced programmers.

© Adam Aspin 2022
A. Aspin, *Pro Data Mashup for Power BI*, https://doi.org/10.1007/978-1-4842-8578-7_14

- The "M" language is very different from many other widely used programming languages.

- Tweaking a step manually can cause havoc to a carefully wrought data load and transform process.

Moreover, the "M" language is so vast that it would require an entire book. Consequently, I have deliberately chosen to provide only the most superficial of introductions here. For greater detail, I suggest that you consult the Microsoft documentation. This is currently available at the following URL:

- `https://docs.microsoft.com/en-us/powerquery-m/`

In this chapter, I am not going to presume that the reader has any in-depth programming knowledge. I will provide a few comparisons with standard programming concepts to assist any readers that have programmed in VBA, C#, or Java. However, rest assured, the intention is to open up new horizons for passionate Power BI Desktop users rather than spiral off into a complex technical universe.

All of this is probably best understood by building on your existing knowledge and explaining how (simply by using the Power Query graphical interface) you have been writing M code already. Then you can extend this knowledge by learning how to tweak existing code, and finally, you will see how to write M code unaided.

What Is the M Language?

I should, nonetheless, begin with a few technical stakes in the ground to explain what the Power Query Formula Language (or M as everyone calls the language now) is and what it can—and cannot—do.

M is a *functional* language. It is certainly *not* designed to perform general-purpose programming. Indeed, battle-hardened programmers will search in vain for coding structures and techniques that are core to other languages.

At the risk of offending programming purists, I prefer to introduce M to beginners as being a functional language in three ways:

- It exists to perform a simple function, which is to load and transform data.

- It is built on a compendium of over 700 built-in functions, each of which is designed to carry out a specific piece of data load and/or transformation logic.

- It exists as a series of one or more functions, each of which computes a set of input values to a single output value.

To complete the whirlwind introduction, you also need to know that

- M is case-sensitive, so you need to be *very* careful when typing in function keywords and variable names.

- M is strongly typed, which means that you must respect the core types of data elements used and convert them to the appropriate type where necessary. M will not do this for you automatically.

- M is built on a set of keywords, operators, and punctuators.

I don't want to get too technical at this juncture. Nonetheless, I hope that a high-level overview will prepare you for some of the approaches that you will learn later in this chapter.

M and the Power Query Editor

The good news about M is that you can already write it. By this, I mean that every example that you followed in the previous 13 chapters wrote one or more lines of M code for you. Indeed, each step in a data load and transformation process that you generated when using the Power Query Editor created M code for you—automatically.

This means several (very positive) things:

- You do not necessarily have to begin writing M code from a blank slate. Often, you can use the Power Query interface to carry out most of the work—and then tweak the code the interface generates to add the final custom elements that you require.

- You do not have to learn over 700 functions to deliver M code as Power Query can find and write many of the appropriate instructions for you.

- The Power Query interface is tightly linked to the way that M code is written. So understanding how to use the interface helps you in understanding what M code is and how it works.

Modifying the Code for a Step

If you feel that you want to delve into the inner reaches of Power Query, you can modify steps in a query by editing the code that is created automatically every time that you add or modify a query step.

To get a quick idea of what can be done:

1. Open Power Query and load the Excel file
   `C:\PowerBiDesktopSamples\BrilliantBritishCars.xlsx`
   (there is only one source table named BaseData). Alternatively,
   load the data from the Power BI report window and click
   Transform Data to open Power Query.

2. Select the column IsDealer and remove it.

3. Click the Removed Columns step in the BaseData query.

4. In the View menu, ensure that the formula bar check box is
   checked. You will see the "M" code in the formula bar under the
   menu ribbon. It will look like that shown in Figure 14-1.

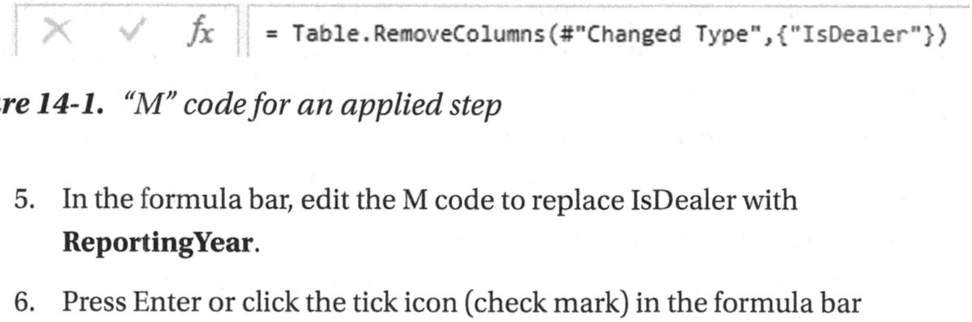

```
×   ✓   fx   | = Table.RemoveColumns(#"Changed Type",{"IsDealer"})
```

Figure 14-1. *"M" code for an applied step*

5. In the formula bar, edit the M code to replace IsDealer with
   ReportingYear.

6. Press Enter or click the tick icon (check mark) in the formula bar
   to confirm your changes.

The step and subsequent data will be updated to reflect your changes.

The modification that you carried out in step 5 effectively means that you are adding back the IsDealer column and removing the ReportingYear column instead. You could have done this using the interface (by clicking the gear cog icon in the Applied Steps list for this step), but the whole point is to understand that both options are available and that the Power Query interface is only generating and modifying M code. So you can modify this code directly, if you prefer.

If you are an Excel power user (as many Power BI aficionados are), then you can be forgiven for thinking that this is similar to Excel Macro development. Indeed, it is in some respects:

- The core code can be recorded (as can be the case for VBA in Excel, M is recorded automatically for Power BI).

- The resulting code can then be modified.

This is, of course, a very simplistic comparison. The two approaches may be similar, but the two languages are vastly different. Yet if this helps as a metaphor to encourage you to move to M development, then so be it.

There are inevitably a series of caveats when modifying the M code for a query step in the formula bar. These include (but are far from restricted to) the following:

- Any error will not only cause the step to fail, it will cause the whole data load and transformation process to fail from the current step onward.

- You need to remember that M is *case-sensitive*, and even the slightest uppercase character in place of a lowercase letter can cause the entire process to fail.

- The use of quotes to define literal elements (such as column names) must be respected.

- M makes lavish use of both parentheses and braces. It can take some practice and understanding of the underlying logic to appreciate their use fully in various contexts.

Fortunately, M will provide fairly clear error messages if (or when) errors creep in. If you enter an erroneous field name, for instance, you could see a message like the one in Figure 14-2.

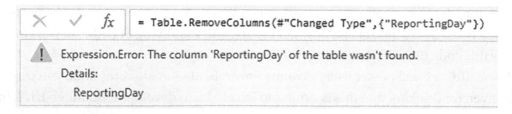

Figure 14-2. *An "M" error message*

I do not want you to feel that modifying M code is difficult or dangerous, however. So to extend the preceding example (and to encourage you to use M), this is what the M code would look like if you extended it to remove two columns, and not just one:

```
= Table.RemoveColumns(#"Changed Type",{"ReportingYear", "IsDealer"})
```

Note   More generally, it is often best to look at the code for existing steps—or create "dummy" code using a sample dataset in parallel—to get an idea of what the M code for a particular function looks like. This could give you ideas of ways to modify the code.

M Expressions

To give you a clearer understanding of what each M "step" contains, Figure 14-3 shows the core structure of a step. However, only the Power Query interface calls this a step. M actually calls this an *expression*. So that is the term I will use from now on.

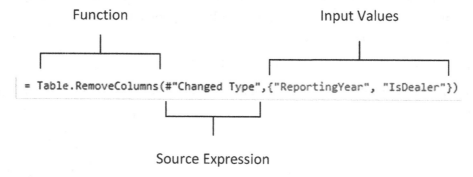

Figure 14-3. *An M expression*

There are several fundamental points that you need to be aware of here:

- Each M *expression* is made up of *functions*. These can be any of the built-in functions (such as the Table.RemoveColumns function used here)—or functions that you have defined (which are explained a little later in this chapter). They can also be calculations or simple logic.

- As you learned in the course of this book so far, data mashup is essentially a series of individual actions (or *steps* as the Power Query interface calls them). These actions are linked in a "chain" where each expression is built on—and refers to—a preceding expression. In Figure 14-3, this specific expression refers to the output of the #"Changed Type" expression that preceded it.

- M expressions can become extremely complex and include multiple functions—rather like advanced Excel formulas. As functions can be nested, this can lead to M expressions becoming complex and dense.

Writing M by Adding Custom Columns

Another way to write certain types of M code is to add custom columns. Although these are known as *custom columns* in Power BI Desktop, they are also known more generically as *derived columns* or *calculated columns*. Although they can do many things, their essential role is to

- Concatenate (or join, if you prefer) existing columns

- Add calculations to the data table

- Extract a specific part of a column

- Add flags to the table based on existing data

The best way to understand these columns is probably to see them in action. You can then extend these principles in your own processes. This can, however, be an excellent starting point to learn basic M coding—albeit limited to a narrowly focused area of data wrangling in M.

Initially, let's perform a column join and create a column named Vehicle, which concatenates the Make and Model columns with a space in between.

1. Open a blank Power BI Desktop file.

2. Connect to the `C:\PowerBiDesktopSamples\` `BrilliantBritishCars.xlsx` data source.

3. Click Transform Data to open the Power Query Editor.

4. In the Add Column ribbon, click Custom Column. The Add Custom Column dialog is displayed.

5. Click the Make column in the column list on the right, and then click the Insert button; `=[Make]` will appear in the Custom column formula box at the left of the dialog.

6. Enter & " " & in the Custom column formula box after `=[Make]`.

7. Click the Model column in the column list on the right, and then click the Insert button.

8. Click inside the New column name box and enter a name for the column. I call it CarType. The dialog will look like Figure 14-4.

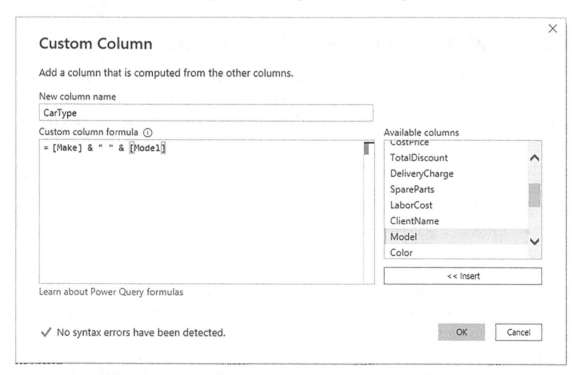

Figure 14-4. *The Custom Column dialog*

9. Click OK. The new column is added to the right of the data table; it contains the results of the formula. Inserted Column appears in the Applied Steps list. The formula bar contains the following formula:

```
= Table.AddColumn(#"Changed Type", "CarType", each [Make]
& " " & [Model])
```

You can always double-click a column to insert it into the Custom column formula box if you prefer. To remove a column, simply delete the column name (including the square brackets) in the Custom column formula box.

Tip   You must always enclose a column name in square brackets.

You can see that this line of M code follows the principles that you have already seen. It uses an M formula (Table.AddColumn) that refers to a previous expression (#"Changed Type") and then applies the code that carries out the expression requirements—in this case, adding a new column that contains basic M code.

Note   The *each* keyword is an M convention to indicate that every record in the column will have the formula applied.

The Advanced Editor

The formula bar is only the initial step to coding in M. Nearly always you will write M code in the Power Query Advanced Editor. There are several fundamental reasons for this:

- The Advanced Editor shows all the expressions that make up an M query.

- It makes understanding the sequencing of events (or steps or expressions if you prefer) much easier.

- The Advanced Editor (like the Formula bar) has built-in IntelliSense. This means that you can see M functions listed as you type.

441

- It has a syntax checker that helps isolate and identify syntax errors.

- It color-codes the M code to make it more readable and comprehensible.

Expressions in the Advanced Editor

The M expressions that you can see individually in the formula bar do not exist in a vacuum. Quite the contrary, they are always part of a coherent sequence of data load, cleansing, and transformation events. This is probably best appreciated if you now take a look at the whole block of M code that was created when you loaded a table from an Excel file previously.

To see the M code, you need to open the Advanced Editor:

1. In the Power Query Home ribbon, click the Advanced Editor button. The Advanced Editor window will open, as shown in Figure 14-5. You can also see the Applied Steps list from the Power Query Editor to help you understand how each step is, in fact, an M expression.

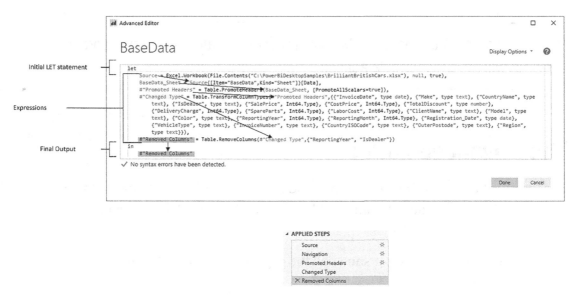

Figure 14-5. *Using the Advanced Editor to edit M code*

This dialog contains the entire structure of the connection and transformation process that you created. It contains the following core elements:

- A sequence of expressions (which are steps)

- A *let* expression that acts as an outer container for sequence of data transformation expressions

- An *in* expression that returns the output of the entire query

If you look at Figure 14-5, you can see several important things about the sequence of expressions that are inside the Let...In block:

- Each expression is named—and you can see its name in the Applied Steps list.

- Each expression refers to another expression (nearly always the previous expression) except for the first one.

- All but the final expression are terminated by a comma.

- An expression can run over several lines of code. It is the final comma that ends the expression in all but the last expression.

- The final expression becomes the output of the query.

Although this is a fairly simple M query, it contains all the essential elements that show how M works. Nearly every M query that you build will reflect these core principles:

- Have a Let...In block

- Contain one or more expressions that contain functions

The Let Statement

The *let* statement is a core element of the M language. It exists to allow a set of values to be evaluated individually where each is assigned to a variable name. These variables form a structured sequence of evaluation processes that are then used in the output expression that follows the *in* statement. You can consider it to be a "unit of processing" in many respects. Let statements can be nested to add greater flexibility.

In most let statements, the sequence of variables will be ordered from top to bottom (as you can see in Figure 14-6) where each named expression refers to, and builds on, the previous one. This is the way that Power Query presents named expressions as steps and is generally the easiest way to write M scripts that are easy to understand. However, it is not, technically, necessary to order the expressions like this as the expressions can be in any order.

Modifying M in the Advanced Editor

As with all things Power BI related, the Advanced Editor is best appreciated through an example. You saw in Chapter 3 how to create and modify connections to data sources. You can also modify connections directly in the "M" language. This assumes that you know and understand the database that you are working with.

1.   Add a new query that connects to a SQL Server database. I am using a SQL instance and database on my PC.

2.   Select this query in the Queries list on the left.

3.   In the Home ribbon, click the Advanced Editor button. The Advanced Editor dialog will appear, as shown in Figure 14-6.

Figure 14-6. *The Advanced Editor dialog to alter a database connection*

4.   Alter any of the following elements:

a.   The server name in the Source line (currently "ADAM03\SQL2017")

b. The database name in the second line (currently Name="PrestigeCars")

c. The schema name in the third line (currently Schema="Data")

d. The table name in the third line (currently Item="Sales")

5. Click Done to confirm any changes and close the Advanced Editor.

This approach really is working without a safety net, and I am showing you more to raise awareness than anything else. However, it does open the door to some far-reaching possibilities if you wish to extend data transformation processes using the "M" language.

Note   You can, of course, click Cancel to ignore any changes that you have made to the M code in the Advanced Editor. Power Query will ask you to confirm that you really want to discard your modifications.

Syntax Checking

If you intend to write and modify M code, you are likely to be using the Advanced Editor—a lot. Consequently, it is certainly worth familiarizing yourself with the help that it can provide. Specifically, its syntax checking can be extremely useful and is entirely automatic.

Suppose that you have (heaven forbid!) made an error in your code. The Advanced Editor could look something like the one in Figure 14-7.

Figure 14-7. Syntax checking in the Advanced Editor

As you can see, in this case, the Advanced Editor no longer displays a check box under the code and the reassuring message "No syntax errors have been detected." Instead, you see an error message, and any errors are underlined in red. Clicking the Show error link will highlight the source of the error by displaying it on a gray background.

Advanced Editor Options

The Advanced Editor also has a few optional settings that you can configure.

1. In the top right of the Advanced Editor dialog, click the pop-up triangle to the right of Display Options. You can see this in Figure 14-8.

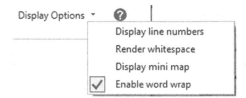

Figure 14-8. *Advanced Editor options*

2. Select Enable word wrap. Any individual M expressions that are too long to fit on a single line in the Advanced Editor will flow onto the next line.

The available options are explained in Table 14-1.

Table 14-1. *Advanced Editor Options*

Option	Description
Display line numbers	Adds line numbers to the left of the code
Render whitespace	Displays whitespace as gray dots
Display mini map	Shows a high-level overview of the code structure on the right of the Advanced Editor, which can be used to scroll to the line of code you want to jump to
Enable word wrap	Allows long lines of code to flow onto the following line

Basic M Functions

The M language is vast—far too vast for anything other than a cursory overview in a single chapter. Nonetheless, to give some structure to the overview, it is worth knowing that there are a few key categories of M functions that you might find useful when beginning to use M.

The following list is not exhaustive by any means but can, hopefully, serve as a starting point for your journey into M functions. The elementary categories are

- Text functions
- Date functions
- Time functions
- DateTime functions
- Logical functions
- Number functions

I am focusing on these categories as they are probably the most easily comprehensible in both their application and their use. Once you have seen some of these functions, we can move on to other functions from the range of those available.

Most of the more elementary M functions can be applied in ways that will probably remind you of their Excel counterparts. For instance, if you want to extend the formula that you used to concatenate the Make and Model columns so that you are only extracting the leftmost three characters from the Make, you can wrap the Make column in the Test.Start() function like this:

```
= Table.AddColumn(#"Changed Type", "CarType", each Text.Start([Make], 3) &
" " & [Model])
```

The result is shown in Figure 14-9.

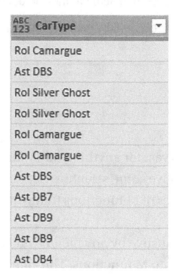

Figure 14-9. *Applying a first text function*

As you can see, wrapping the Make column inside this particular text function has added an extra layer of data transformation to the expression. Indeed, if you are familiar with SQL or Excel functions, you may have met the LEFT() function that these use. Well, Text.Start() in M is very similar.

Text Functions

Rather than take you step by step through every possible example of text functions, I prefer to show you some of the more useful text functions (at least, in my experience). These code snippets are given in Table 14-2, where you will doubtless recognize much of the functionality that you have been accessing up until now through the Power Query user interface.

You can see also that many M functions require more than one element inside the parentheses. These elements are called parameters–and adding the required number of parameters in the correct order is *vital* to getting M to work correctly.

Table 14-2. *Text Function Examples*

Output	Code Snippet	Description
Left	`Text.Start([Make],3)`	Returns the first three characters from the Make column
Right	`Text.End([Make],3)`	Returns the last three characters from the Make column
Up to a specific character	`Text.Start([Make],Text.PositionOf([Make]," "))`	Returns the leftmost characters up to the first space
Up to a delimiter	`Text.BeforeDelimiter([InvoiceNumber], "-" ,2`	Returns the text before the third hyphen
Text length	`Text.Length([Make])`	Finds the length of a text
Extract a substring	`Text.Range([Make], 2, 3)`	Extracts a specific number of characters from a text—starting at a specified position
Remove a substring	`Text.RemoveRange([Make], 2, 3)`	Removes a specific number of characters from a text—starting at a specified position
Replace a text	`Text.Replace([Make], "o", "a")`	Replaces all the o characters with an a in the text or column
Trim spaces	`Text.Trim([Make])`	Removes leading and trailing spaces in the text or column
Convert to uppercase	`Text.Upper([Make])`	Converts the text or column to uppercase
Convert to lowercase	`Text.Lower([Make])`	Converts the text or column to lowercase
Add initial capitals	`Text.Proper([Make])`	Adds initial capitals to each word of the text or column

---

Note   You have probably noticed if you looked closely at these functions that any numeric parameters are zero based. So to define the third hyphen when splitting text in a column, you would use *2*, not *3*.

Table 14-2 is only a subset of the available text functions in M. If you want to see the complete list, it is on the Microsoft website at `https://docs.microsoft.com/en-us/powerquery-m/text-functions`.

There are, inevitably, many more text functions available in M. However, the aim is not to drown the reader in technicalities, but to make you aware of both the way that M works and what is possible.

M or DAX?

Once you also learn the DAX language to create calculations in Power BI, you may well wonder why you carry out operations like this in Power Query when you can do virtually the same thing in the data model. Well, it is true that there is some overlap; so you have the choice of which to use. You can perform certain operations at multiple stages in the data preparation and analysis process. It all depends on how you are using the data and with what tool you are carrying out the analyses. A good rule of thumb is that transformations that are used to prepare the data should be pushed up to Power Query where possible.

Number Functions

To extend your knowledge, Table 14-3 shows a few of the available number functions in M. Here, I have concentrated on showing you some of the numeric type conversions as well as the core calculation functions.

Table 14-3. *Number Function Examples*

Output	Code Snippet	Description
Returns an 8-bit integer	`Int8.From("25")`	Converts the text or number to an 8-bit integer
Returns a 16-bit integer	`Int16.From("2500")`	Converts the text or number to a 16-bit integer
Returns a 32-bit integer	`Int32.From("250,000")`	Converts the text or number to a 32-bit integer
Returns a 64-bit integer	`Int64.From("2500000000")`	Converts the text or number to a 64-bit integer
Returns a decimal number	`Decimal.From("2500.123")`	Converts the text or number to a decimal
Returns a Double number value from the given value	`Double.From("2500")`	Converts the text or number to a double-precision floating-point number
Takes a text as the source and converts to a numeric value	`Number.FromText("2500")`	Converts the text to a number
Rounds a number	`Number.Round(5000, 0)`	Rounds the number up or down to the number of decimals (or tens, hundreds, etc., if you use a negative second parameter)
Rounds a number up	`Number.RoundUp(5020, -2)`	Rounds the number up to the number of decimals (or tens, hundreds, etc., if you use a negative second parameter)
Rounds a number down	`Number.RoundDown(100.01235, 2)`	Rounds the number down to the number of decimals (or tens, hundreds, etc., if you use a negative second parameter)
Removes the sign	`Number.Abs(-50)`	Returns the absolute value of the number
Raises to a power	`Number.Power(10, 4)`	Returns the value of the first parameter to the power of the second

(continued)

Table 14-3. (*continued*)

Output	Code Snippet	Description
Modulo	Number.Mod(5, 2)	Returns the remainder resulting from the integer division of number by divisor
Indicates the sign of a number	Number.Sign(-5)	Returns 1 if the number is a positive number, -1 if it is a negative number, and 0 if it is zero
Gives the square root	Number.Sqrt(4)	Returns the square root of the number

Table 14-3 is only a minor subset of the vast range of number functions that are available in M. If you want to see the complete list, it is on the Microsoft website at `https://docs.microsoft.com/en-us/powerquery-m/number-functions`.

Date Functions

M has many date functions. Table 14-4 contains a potentially useful sample of the available functions.

Table 14-4. *Date Function Examples*

Output	Code Snippet	Description
Day	Date.Day(Date. FromText("25/07/2020"))	Returns the number of the day of the week from a date
Month	Date.Month(Date. FromText("25/07/2022"))	Returns the number of the month from a date
Year	Date.Year(Date. FromText("25/07/2022"))	Returns the year from a date
Day of week	Date.DayOfWeek(Date. FromText("25/07/2022"))	Returns the day of the week from a date (0 being Monday)
Name of weekday	Date.DayOfWeekName(Date. FromText("25/07/2022"))	Returns the weekday name from a date
First day of month	Date.StartOfMonth(Date. FromText("25/07/2022"))	Returns the first day of the month from a date
Last day of month	Date.EndOfMonth(Date. FromText("25/07/2022"))	Returns the last day of the month from a date
First day of year	Date.StartOfYear(Date. FromText("25/07/2022"))	Returns the first day of the year from a date
Last day of year	Date.EndOfYear(Date. FromText("25/07/2022"))	Returns the last day of the year from a date
Day of year	Date.DayOfYear(Date. FromText("25/07/2022"))	Returns the day of the year from a date
Week of year	Date.WeekOfYear(Date. FromText("25/07/2022"))	Returns the week of the year from a date
Quarter	Date.QuarterOfYear(Date. FromText("25/07/2022"))	Returns the number of the quarter from a date
First day of quarter	Date.StartOfQuarter(Date. FromText("25/07/2022"))	Returns the first day of the quarter from a date
Last day of quarter	Date.EndOfQuarter(Date. FromText("25/07/2022"))	Returns the last day of the quarter from a date

Table 14-4 is only a subset of the available date functions in M. If you want to see the complete list, it is on the Microsoft website at `https://docs.microsoft.com/en-us/powerquery-m/date-functions`.

Time Functions

M also has many time functions. Table 14-5 contains a potentially useful sample of the available functions.

Table 14-5. *Time Function Examples*

Output	Code Snippet	Description
Hour	`Time.Hour(#time(14, 30, 00))`	Returns the hour from a time
Minute	`Time.Minute(#time(14, 30, 00))`	Returns the minute from a time
Second	`Time.Second(#time(14, 30, 10))`	Returns the second from a time
Time from fraction	`Time.From(0.5)`	Returns the time from a fraction of the day

Table 14-5 is only a subset of the available time functions in M. If you want to see the complete list, it is on the Microsoft website at `https://docs.microsoft.com/en-us/powerquery-m/time-functions`.

Equally, as they are so similar to the date and time functions, I have not shown here the datetime functions and the datetimezone functions. These are also available on the Microsoft website.

Duration Functions

M can also extract durations—in days, hours, minutes, and seconds. Table 14-6 shows some of the basic duration functions.

Table 14-6. *Duration Function Examples*

Output	Code Snippet	Description
Days	`Duration.Days(#duration(10, 15, 55, 20))`	Duration in days
Hours	`Duration.Hours(#duration(10, 15, 55, 20))`	Duration in hours
Minutes	`Duration.Minutes(#duration(10, 15, 55, 20))`	Duration in minutes
Seconds	`Duration.Seconds(#duration(10, 15, 55, 20))`	Duration in seconds

Table 14-6 is nearly all the available duration functions in M. If you want to see the remaining few functions, they are on the Microsoft website at `https://docs.microsoft.com/en-us/powerquery-m/duration-functions`.

M Concepts

The time has now come to "remove the stabilizers" from the bicycle and learn how to cycle unaided. This means, firstly, becoming acquainted with several M structural concepts.

This means moving on from the "starter" functions that you can use to modify the contents of the data to creating and modifying data structures themselves. M is essentially focused on loading and presenting tabular data structures, so tables of data are an essential data structure. However, there are other data structures that it can manipulate—and that you have seen in passing in previous chapters. In this chapter, then, we will look at the three core data structures. Collectively, these are classified as *structured* values—as opposed to the *primitive* values such as text, number, or date and time. Some of these are

- Lists

- Records

- Tables

However, before delving into these structured data elements, you need to understand two fundamental aspects of the M language. These are

- Data types

- M values (also referred to as variables or identifiers)

So without further ado, let's continue your journey into M.

M Data Types

If you are creating your own lists, records, and tables, then it will help to know the basics about data types in M.

When beginning to use M, you need to remember that primitive data values must always be one of the following types:

- Number

- Text

- Date

- Time

- DateTime

- DateTimeZone

- Duration

- Logical (Boolean if you prefer)

- Binary

- Null

There are other types such as function, any, or anynonnull, but we will not be covering them in this book.

All data types expect to be entered in a specific way. Indeed, you must enter data in the way shown in Table 14-7 to avoid errors in your M code.

Table 14-7. *Data Type Entry*

Data Type	Code Snippet	Comments
Number	100 0.12345 2.4125E8 2.4125E-8	Do not use formatting such as thousands separators or monetary symbols.
Text	"Calidra Power BI Training"	Always enclose in double quotes. Use two double quotes to enter the actual quoted text.
Date	#date(2022,12,25)	Dates must be year, month, and day in the #date() function.
Time	#time(15,55,20)	Times must be hour, minute, and second in the #time() function.
DateTime	#datetime(2022,12,25,15,55,20)	Datetimes must be year, month, day, hour, minute, and second in the #datetime() function.
DateTimeZone	#datetimezone(2022,12,25,15,55,20,-5,-30)	Datetimezones must be year, month, day, hour, minute, second, hour offset, and minute offset in the #datetimezone() function.
Duration	#duration(0,1,0,0)	Days, hours, minutes, and seconds comma-separated inside the #duration() function.
Logical	true	true or false in lowercase.

M Values

Before practicing some actual coding in M, you really need to know a few fundamentals concerning M values:

- Values are the output of expressions.

- Values are also variables.

- The names of values are case-sensitive.

- If the value name contains spaces or restricted characters, they must be wrapped in #"" (pound sign followed by double quotes).

This fourth bullet point clearly begs the question "what is a restricted character?" The simple answer—that avoids memorizing lists of glyphs—is "anything not alphanumeric."

Note   The Power Query Editor interface makes the steps (which are the values returned by an expression) more readable by adding spaces wherever possible. Consequently, these values always appear in the M code as #"Step Name".

Defining Your Own Variables in M

As the values returned by any expression are also variables, it follows that defining your own variables in M is breathtakingly simple. All you have to do is to enter a variable name (with the pound sign and in quotes if it contains spaces or restricted characters), an equals sign, and the variable definition.

As a really simple example, take a look at Figure 14-10. This M script defines the three parameters required for the List.Numbers() function that you will see in Table 14-8 and then uses the variables inside the function. The List.Numbers() function is used to create a list of numbers that you can use inside other M processes.

Figure 14-10. *User variables in M*

There are only a few things to remember when defining your own variables:

- They respect the same naming convention as output values in M.

- A variable can be referenced inside the subexpression where it is defined and any expressions that contain the subexpression.

- A variable can be a simple value or a calculation that returns a value.

Writing M Queries

Before actually writing M, you need to know how and where to write your code. Suppose that you need an environment to practice the examples in the remainder of this chapter:

1. Open a new, blank Power BI Desktop file.

2. In the Home ribbon, click Get Data ➤ Blank Query. The Power Query Editor will open.

3. Click Advanced Editor. An Advanced Editor dialog will open containing only an outer let expression, as shown in Figure 14-11.

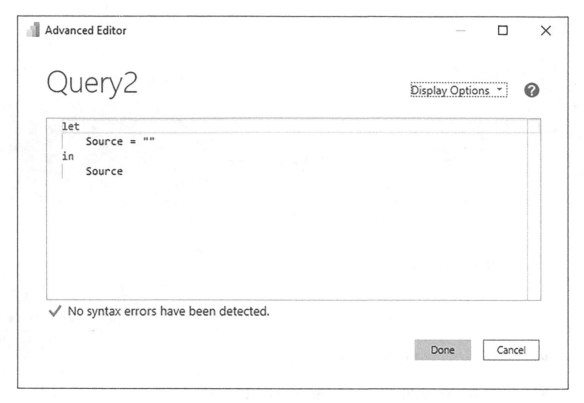

Figure 14-11. *Preparing the Advanced Editor to write M code*

Note   Technically, a let clause is not required in M. You can simply enter an expression. However, I prefer to write M "by the book"—at least to begin with.

M Autocomplete

Fortunately, you will not have to memorize 700-odd M functions and type them into a text editor to write M code. This is because Power Query will assist you using Autocomplete.

This means that (a bit like when writing Excel formulas) Power Query will suggest both functions and any variables you have created when you write M scripts. You can see this in Figure 14-12.

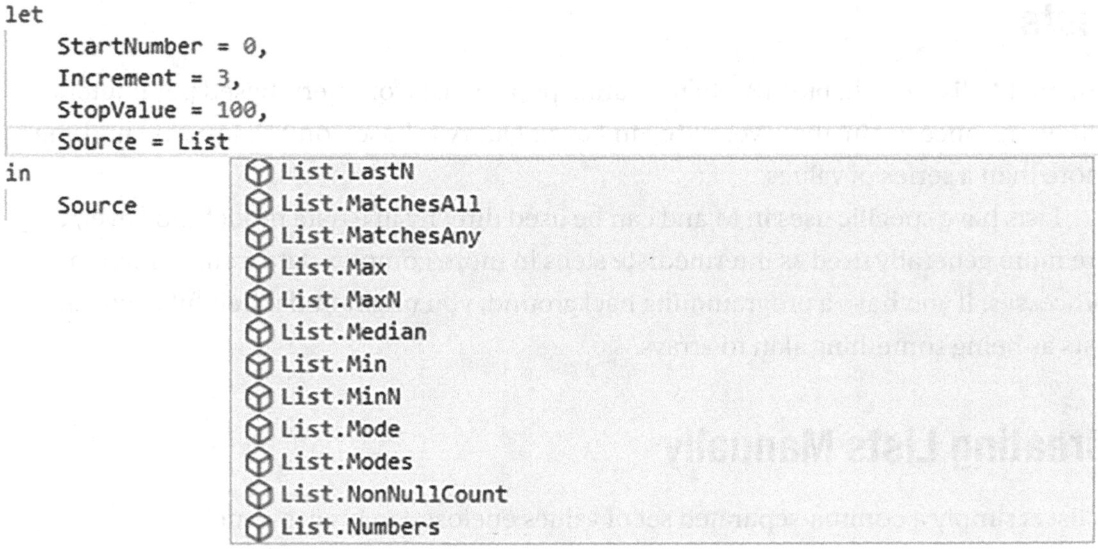

```
let
    StartNumber = 0,
    Increment = 3,
    StopValue = 100,
    Source = List
in
    Source
```

🔷	List.LastN
🔷	List.MatchesAll
🔷	List.MatchesAny
🔷	List.Max
🔷	List.MaxN
🔷	List.Median
🔷	List.Min
🔷	List.MinN
🔷	List.Mode
🔷	List.Modes
🔷	List.NonNullCount
🔷	List.Numbers

Figure 14-12. *Autocomplete in the Advanced Editor when writing M code*

The Advanced Editor will further assist you by reminding you of the parameters of each M function that you apply. You can see an example of this in Figure 14-13. This information appears when you have selected a function and are in the process of adding any required parameters.

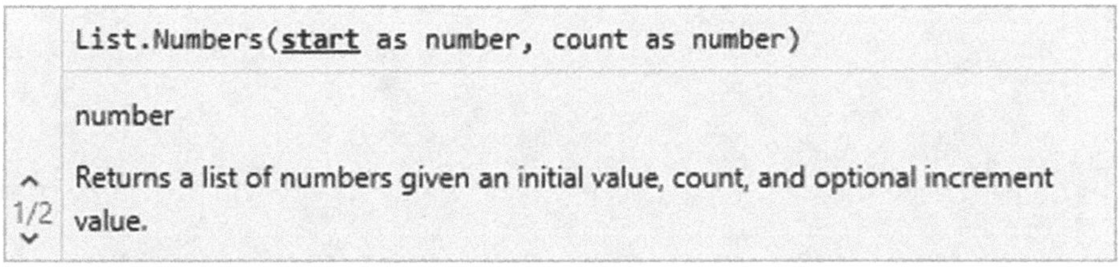

```
List.Numbers(start as number, count as number)

number

Returns a list of numbers given an initial value, count, and optional increment
1/2  value.
```

Figure 14-13. *M function hints*

The Advanced Editor will try and help you in other ways as well. It will, for instance:

- Complete matching sets of curly braces by adding the right brace when you type in a left brace

- Complete matching sets of parentheses by adding the right parenthesis when you type in a left parenthesis

Lists

You met M lists in Chapter 13 when creating pop-up lists for query-based parameters (showing, once again, that everything in Power Query is based on M). Lists are nothing more than a series of values.

Lists have specific uses in M and can be used directly in a data model. However, they are more generally used as intermediate steps in more complex data transformation processes. If you have a programming background, you might find it helpful to consider lists as being something akin to arrays.

Creating Lists Manually

A list is simply a comma-separated set of values enclosed in braces—such as

{1,2,3}

Once integrated into the structure of an M query, it could look like the example shown in Figure 14-14.

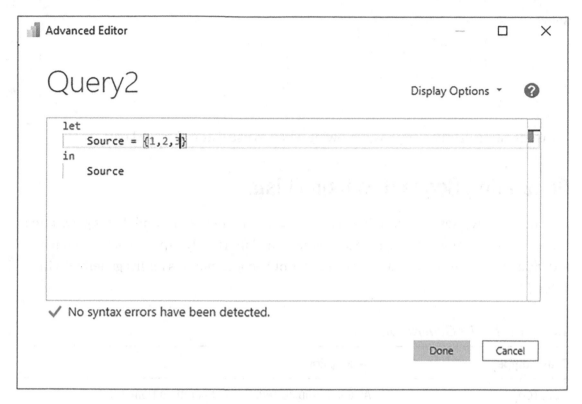

Figure 14-14. *A list in the Advanced Editor*

Once created—either as an intermediate step in a query or as the final output of the query—the list can be used by anything that requires a list as its input. Indeed, if you click the Done button for the preceding example, the Power Query Editor will display this piece of M code as a functioning list—exactly like the one that you created in Chapter 13. So you can now create custom lists for parameters (for instance) quickly and easily.

There is some technical information that you need to know about lists:

- Lists are unlimited in size.

- Lists can contain data of the same type (i.e., all elements are numeric values, dates, or texts, for instance)—or the data can be of different types.

- Lists can be empty—that is, composed of a pair of empty braces.

- Lists can be entered horizontally or vertically. That is, the list shown earlier could have been typed in as

```
Source = {
          1,
          2,
          3
        }
```

Generating lists is really easy; knowing when to use lists is the hard part.

Generating Sequences Using Lists

Lists have many uses in M, but there is one area where they shine, and that is generating sequences of numbers, dates, or texts. Rather than laboriously explain each approach individually, I have collated a set of examples of M code snippets for list generation in Table 14-8.

Table 14-8. *List Generation*

Code Snippet	Description
{1..100}	An uninterrupted sequence of numbers from 1 to 100, inclusive
{1..100, 201..400}	An uninterrupted sequence of numbers from 1 to 100, then from 201 to 400
List.Numbers(0, 100, 5)	Starting at zero increments by 5 until 100 numbers have been written
{"A".."Z"}	The uppercase letters A through Z
List.Dates(#date(2022, 1, 1), 366, #duration(1, 0, 0, 0))	Each individual day for the year 2022—starting on January 1, 2022, 366 days (expressed as a duration in days) are added
List.Times(#time(1, 0, 0), 24, #duration(0, 1, 0, 0))	Each hour in the day starting with 1 AM

Accessing Values from a List

If you move to more advanced M coding, you may well want to refer to a value from a list in your M script. At its simplest, this is done using *positional references*. Here is a short piece of M code that does just this:

```
let
    Start = {"George","Bill","George W.", "Barack", "Donald"},
    source = Start{3}
in
    source
```

The output of this code snippet is the fourth element in the list—making the point that lists in M are *zero based*. That is, the first element in a list is the element 0.

List Functions

There are many dozens of list functions available in M. Far too many to go through in detail here. So to give you an idea of some of the possible ways that you can manipulate lists, take a look at Table 14-9. All of them use the very simple list that you saw previously.

Table 14-9. *List Functions*

Output	Code Snippet	Description
First value	`List.First(MyList)`	Returns the first element in a list
Last value	`List.Last(MyList)`	Returns the last element in a list
Sort list values	`List.Sort(MyList)`	Sorts the values in a list with optional comparison criteria
Extract range	`List.Range(MyList, 4)`	Extracts a range of values from a list (takes all elements from the 5th element until the end)
Return value(s)	`List.Select(MyList , each _ ="Adam")`	Returns the elements from a list that match a criterion
Generate a list	`List.Generate(() => 10, each _ > 0, each _ - 1)`	Creates a list of sequential values
Aggregate values	`List.Sum(MyList)`	Aggregates the numeric values in a list
Replace values	`List.ReplaceMatchingItems(MyList, {{"Joe", "Fred"}})`	Replaces a range of values in a list
Convert to list	`Table.Column(MyList)`	Returns a column from a table as a list

Table 14-9 is only a small subset of the available list functions in M. If you want to see the full range of functions, it is on the Microsoft website at `https://docs.microsoft.com/en-us/powerquery-m/list-functions`.

Records

If lists can be considered as columns of data that you can use in your M code, *records* are rows of data. You might well find yourself needing to define records when creating more complex data transformation routines in M.

At its simplest, here is a sample record created in M:

```
let
    Source = [Surname = "Aspin", FirstName = "Adam"]
in
    Source
```

If you need to access the data in a record, you append the record variable name with the field name in square brackets, like this:

```
let
    Source = [Surname = "Aspin", FirstName = "Adam"],
    Output = Source[Surname]
in
    Output
```

There are a few record functions that you may find useful. These are outlined in Table 14-10.

Table 14-10. *Record Functions*

Output	Code Snippet	Description
Add field	Record.AddField(Source, "MiddleName", "James")	Adds a field to a record
Remove field	Record.RemoveFields(Source, {"Surname"})	Removes a field from a record
Rename fields	Record.RenameFields(Source, {{"Surname", "LastName"}})	Renames a field in a record
Output field	Record.Field(Source, "Surname"))	Returns the value of the specified Source, {{"Surname", "LastName"}} field in the record
Count	Record.FieldCount(Source)	Returns the number of fields in a record

Table 14-10 is only a subset of the available record functions in M. If you want to see the complete list, it is on the Microsoft website at https://docs.microsoft.com/en-us/powerquery-m/record-functions.

Tables

The final structured data type that you could well employ in M code is the *table* type. As you might expect in a language that exists to load, cleanse, and shape tabular data, the table data type is fundamental to M.

If you decide to create your own tables manually in M, then you will need to include, at a minimum, the following structural elements:

- The #table() function

- A set of column/field headers where each field name is enclosed in double quotes and the set of field names is wrapped in braces

- Individual rows of data, each enclosed in braces and comma separated, where the collection of rows is also wrapped in braces

A very simple example of a hand-coded table could look like this:

```
#table(
      {"Surname", "FirstName"},
      {
          {"Smith","Emmanuel"},
          {"Jones","Angela"},
          {"Brown","Boris"}
      }
    )
```

However, the weakness with this approach is that there are no type definitions for the fields. Consequently, a much more robust approach would be to extend the table like this:

```
#table(
      type table
              [
                    #"Surname" = text,
                    #"FirstName" = text
              ],
          {
              {"Smith"," Emmanuel "},
              {"Jones","Angela"},
```

```
            {"Brown","Boris"}
        }
    )
```

Note   The data type keywords that you specify to define the required data type were outlined earlier in this chapter.

There are many table functions available in M. I have outlined a few of the more useful ones in Table 14-11.

Table 14-11. *Table Functions*

Output	Code Snippet	Description
Merge tables	`Table.Combine()`	Merges tables of similar or different structures
Number of records	`Table.RowCount()`	Returns the number of rows in a table
First	`Table.First()`	Returns the first row in a table
Last	`Table.Last()`	Returns the last row in a table
Find rows	`Table.FindText()`	Returns the rows in the table that contain the required text
Insert rows	`Table.InsertRows()`	Inserts rows in a table
Output rows	`Table.Range()`	Outputs selected rows
Delete rows	`Table.DeleteRows()`	Deletes rows in a table
Select columns	`Table.SelectColumns()`	Outputs selected columns

Table 14-11 is, as you can probably imagine, only a tiny subset of the available table functions in M. If you want to see the complete list, it is on the Microsoft website at `https://docs.microsoft.com/en-us/powerquery-m/table-functions`.

Other Function Areas

As I mentioned previously, M is a vast subject that could fill an entire (and very large) book. We have taken a rapid overview of some of the core concepts and functions, but there is much that remains to be learned if you wish to master M. If you are really interested in learning more, then I suggest that you search the Microsoft documentation for the elements outlined in Table 14-12 to further your knowledge.

Table 14-12. *Other Function Areas*

Function Area	Description
Accessing data functions	Accesses data and returns table values
Binary functions	Accesses binary data
Combiner functions	Used by other library functions that merge values to apply row-by-row logic
DateTime functions	Functions applied to datetime data
DateTimeZone functions	Functions applied to datetime data with time zone information
Expression functions	M code that was used for expressions
Line functions	Converts data to or from lists of values
Replacer functions	Used by other functions in the library to replace a given value in a structure
Splitter functions	Splits values into subelements
Type functions	Returns M types
URI functions	Handles URLs and URIs
Value functions	Handles M values

Custom Functions in M

M also allows you to write custom functions that can carry out highly specific tasks repeatedly.

As an example of a very simple custom function, try adding the following code snippet to a new, blank query:

```
let DiscountAnalysis =
                (Discount as number) =>
                                if Discount < 10 then
                                "Poor" else "Excellent" in
                                DiscountAnalysis
```

You can see the Advanced Editor that contains this user-defined function in Figure 14-15.

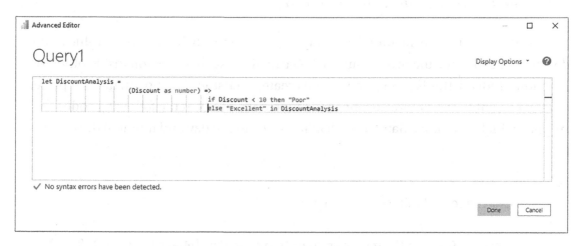

Figure 14-15. *Creating a user-defined function in M*

When you close the Advanced Editor, you will see that this query has been recognized as being an M function and appears as such in the list of queries. You can see this in Figure 14-16 (where I have renamed it to "DiscountAnalysis").

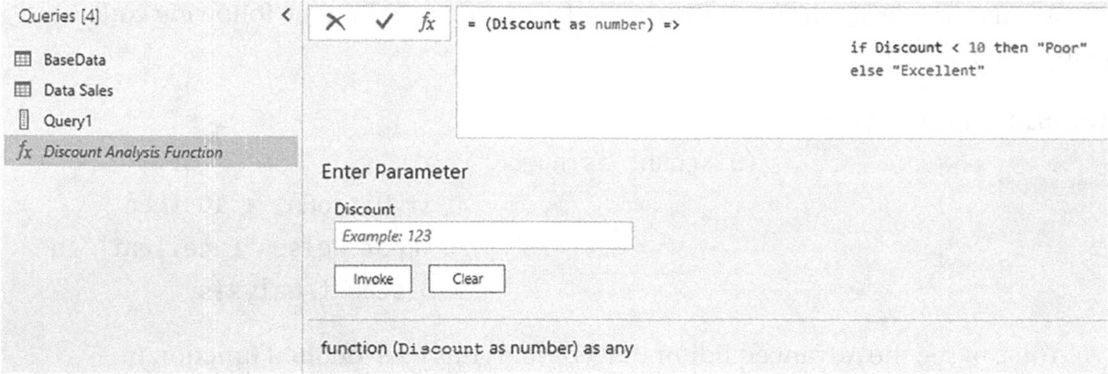

Figure 14-16. *User-defined functions in M*

You can now invoke the function at any time interactively by entering a value as the discount and clicking the Invoke button. You can also use this function inside other M functions. Indeed, this is probably why you created a custom function in the first place.

If you want to see a more advanced function, take a look at the following code snippet, which pads out a date to add leading zeros to the day and month if these are required:

```
let
    FormatDate = (InDate as date) =>
let
    Source = Text.PadStart(Text.From(Date.Month(InDate)),2,"0") & "/"
    & Text.PadStart(Text.From(Date.Day(InDate)),2,"0") & "/" &Text.
    From(Date.Year(InDate))
in
    Source
in
    FormatDate
```

Adding Comments to M Code

Complex M code can be extremely dense. So you will likely need ways of remembering why you created a process when you return to it weeks or months later.

One simple way to make your own life easier is to add comments to M code. You can do this both for code that you have written and queries that have been generated automatically.

There are two ways to add comments.

Single-Line Comments

To comment a single line (which you can do either at the start of the line or partway through the line), simply add two forward slashes—like this:

```
//This is a comment
```

Everything from the two slashes until the end of the line will be considered to be a comment and will not be evaluated by M.

Multiline Comments

Multiline comments can cover several lines—or even part of a line. They cover all the text that is enclosed in /* ... */.

```
/* This is a comment
Over
Several lines */
```

Everything inside the /* ... */ will be considered to be a comment and will not be evaluated by M.

Conclusion

This chapter finishes your introduction to data mashup with Power Query. In this chapter, you learned the basics of the M language that underpin everything that you learned in this book up until now.

You began by seeing how you can use the Power Query Editor interface to assist you in writing short snippets of M code. Then you moved on to discovering the fundamental M concepts such as expressions, variables, and values. Finally, you learned about data types in M and the more complex data types such as lists and tables that underlie complex data transformations. This involved learning to use the Advanced Editor to write and validate your code.

In this chapter and the 13 previous chapters, you have seen essentially a three-stage process: first, you find the data, then you load it into Power Query, and from there, you cleanse and modify it. The techniques that you can use are simple but powerful and can range from changing a data type to merging multiple data tables. Now that your data is prepared and ready for use, you can add it to the Power BI Desktop data model and start creating your Power BI Desktop dashboards.

This brings us to the end of this book on loading and transforming data with Power Query in Power BI Desktop. You can now take the data that you have found, cleansed, and loaded and use it to create powerful and eye-catching dashboards.

It only remains for me to wish you "Good Luck" on your journey with Power BI and Power Query. I sincerely hope that you will have as much fun using Power Query as I had writing this book.

APPENDIX A

Sample Data

Sample Data

If you wish to follow the examples used in this book—and I hope you will—you will need some sample data to work with. All the files referenced in this book are available for download and can easily be installed on your local PC. This appendix explains where to obtain the sample files, how to install them, and what they are used for.

Downloading the Sample Data

The sample files used in this book are currently available on the Apress site. You can access them as follows:

1. In your web browser, navigate to the following URL:
   `github.com/apress/pro-data-mashup-with-powerbi`.

2. Click the button Download Source Code. This will take you to the GitHub page for the source code for this book.

3. Click Clone or Download ➤ Download Zip and download the file PowerBIDesktopSamples.zip.

© Adam Aspin 2022
A. Aspin, *Pro Data Mashup for Power BI*, https://doi.org/10.1007/978-1-4842-8578-7

You will then need to extract the files and directories from the zip file. How you do this will depend on which software you are using to handle zipped files. If you are not using any third-party software, then one way to do this is as follows:

1. Create a directory named `C:\PowerBIDesktopSamples`.

2. In the Windows Explorer navigation pane, click the file PowerBIDesktopSamples.zip.

3. Select all the files and folders that it contains.

4. Copy them to the folder that you created in step 1.

Index

© Adam Aspin 2022
A. Aspin, *Pro Data Mashup for Power BI*, https://doi.org/10.1007/978-1-4842-8578-7

Printed in the United States
by Baker & Taylor Publisher Services